Creative Thinking Activities for Earth Science

Hundreds of Mind Teasers, Puzzles, Riddles, and Word Challenges to Reinforce Major Science Concepts

Written by Robert G. Hoehn

Illustrated by Corbin Hillam

Dedication

To William Latimer, fellow educator at Roseville High School, for your long-lasting friendship and ability to make science come alive with your excellent sense of humor.

Imprint Manager: Kristin Eclov
Editor: Karen P. Hall
Inside Design: Anthony D. Paular
Cover Design: Stephanie Berman

Good Apple
A Division of Frank Schaffer Publications, Inc.
23740 Hawthorne Boulevard
Torrance, CA 90505

GA1691
ISBN 0-7682-0105-5

Table of Contents

Table of Contents

Table of Contents

Table of Contents

CHAPTER I

Matter

Matter is anything on Earth that takes up space—anything you can see, hear, feel, touch, or taste. All matter is made up of atoms and elementary particles, and exists ordinarily as a solid, liquid, or gas.

MIND STUMPERS

1. Matter Chatter

What makes up matter? Does it really matter? Yes, it matters so.
Well, if it does, then use five letters in the words *matter so* and give the problem a go!
Hint: Choose five letters that spell out the name of the particles that make up matter.

2. Sketch Test

What does the following illustration show?

water ———— **one phase to another** ————→

3. King Fling

What is a word that rhymes with *king* and has the same spelling as the name of a liquid and solid?

4. Double Up

Two or more elements may combine to form a compound. How many elements are in the following compound?

COmpound

5. What's in an "E"?

Look at the 111 elements in a periodic table. Examine each of the chemical symbols. Count the number of e's in *the elements*. How many are there?

6. A Bit of Magic

Perspiration, or *sweat*, is a liquid released from the skin to cool off the body when it gets overheated. How could sweat be changed into a solid that warms the body?
Hint: Add a little "letter magic."

THINK OF A WAY

7. Bye, Bye, Baby
Some matter, especially liquids, may be heated in an evaporating dish. Think of a tricky way to illustrate an evaporating dish.

8. What Goes Where?
Think of a way to show a mixture by using four different letters and three arrows.

9. Can It Be Done?
Think of a way to make gold out of lead by using the numbers 207.2 and *10.2*.

10. Liquid Mountains
Think of a way to turn mountains into a liquid by changing one letter.

11. Only One
Think of a way to create a single element from seven letters.

12. A Bit Rusty
Think of a way to create the earth's crust by using the following letters:

C, Fe, O, O, Fe, O

Hint: What is a common name for the compound Fe_2O_3?

13. Combo
Think of a way to get a mixture from matter by changing one letter.

14. Solid Partners
Think of a way to make a mixture of iron and sand by using eight letters.

15. Think Sink
Ice, wood, and cork are examples of matter known to float on water. Think of a way to sink cork by simply rearranging its letters.

THINK OF A WAY

16. Dry as Ice
A compound is a substance composed of two or more different elements that are chemically united. Think of a way to use *COmpOund* to make another compound capable of extinguishing fires.

17. Phase Change
Matter can change from one phase to another. Think of a way to form a shape from one phase.

18. Letter Binge
Think of a way to show physical change by using eight different letters.

19. Bumper Cars
Think of a way, without drawing a picture, to show scattered atoms in *constant motion*.

20. Amorphous
A gas has no definite shape. Think of a way to illustrate this concept by using *hydrogen*.

21. Slow Go
There are six inert gases: helium, neon, argon, krypton, xenon, and radon. Think of a way to highlight one of these gases in the word *inert*.

22. Negative Thinking
Think of a way to express *electrons* by using 32 negative signs.

23. Metal Detector
Think of a way to create metal from nonmetal.

THINK OF A WAY

24. More Metal
Think of a way to form matter as a metal in *motion*.

25. Still More Metal
Think of a way to form matter as two metals in *motion*.

26. Where It Happens
The *nucleus* is the central core of an atom. Think of a way to fit two of them into the following spaces:

27. Enough Energy
Think of a way to show three different energy levels with three *n*'s, three *g*'s, three *r*'s, six *e*'s, and three *y*'s.

28. Water Wonder
Think of a way to use four different elements to spell the name of a water-dwelling animal that has gills, fins, and scales.

29. Early Mixers
Alchemists were chemists in the Middle Ages. They studied the nature of matter. According to history, alchemists were men. Yet it can be shown that only 20 percent of alchemists fit the male gender. Think of a way to demonstrate this.

30. Used Up
Think of a way to use all the letters in the word *alchemist* to form two solids: one gas, and one liquid.

31. Group Effort
Think of a way to show how two alchemists working "across" from each other in the laboratory can be a real "plus."

PUZZLES AND PROBLEMS

❀ ❀

32. All That's Left

The dark brown-to-black fossil remains of plants is a mineral that contains much carbon. Solve the puzzle to discover the name of this substance.

Directions: Darken all spaces in the puzzle that contain either the chemical symbol or the atomic number of carbon. The darkened spaces will reveal three letters. Use these letters and the letter *a* to spell out the name of the substance.

N	S	7	13	H	8	14	O
5	C	6	C	5	C	B	5
B	6	G	W	9	6	K	3
K	C	17	B	H	C	U	L
T	6	C	6	8	6	C	6
8	W	S	V	K	S	12	A
S	A	8	6	C	C	T	4
O	N	S	C	7	C	3	H
2	4	3	6	V	6	K	9
8	3	N	6	C	C	13	W

33. We're Surrounded

Matter is anything that takes up space and has mass. Read the following clues, and then use the letters in the word *matter* to spell the name of each "mystery matter." You may use letters more than once.

a. Rodent: __ __ __

b. Male sheep: __ __ __

c. Mature female horse: __ __ __ __

d. Box-like wagon running on rails: __ __ __ __

e. Device that measures the flow of water, gas, or electricity:

__ __ __ __ __

f. Floor covering: __ __ __

g. The upper limb of a human: __ __ __

h. An object that holds a golf ball: __ __ __

i. Beverage: __ __ __

j. Substance that lubricates the eyeball: __ __ __ __

GA1691 Earth Science © Good Apple

PUZZLES AND PROBLEMS

❀❀❀❀❀❀❀❀❀❀❀❀❀❀❀❀❀❀❀❀❀❀❀❀❀❀❀❀❀❀❀❀❀❀❀❀

34. Show Yourself

Part A: What phase of matter is hiding in the puzzle? Lightly darken the words in the puzzle that rhyme with *phase* to reveal the answer.

b$_5$	l	a	z	e$_6$	c$_{24}$	r	a	z	e$_{12}$
d$_1$	a	y	s	g	h$_{12}$	a$_8$	z	e	
m$_{13}$	a	i$_3$	z	e	a	p	a	y	s
	s	t$_7$	a	y	s$_{20}$	s	r$_9$	a	y	s$_{17}$
w	a	y	s$_4$	p	r$_{10}$	a	i	s$_{14}$	e
	m	a$_{16}$	y	o	n	n	a$_{30}$	i	s$_2$	e

Part B: There is another hidden word in the puzzle above—an example of the mystery phase from Part A. Answer the following questions and find the number values in the puzzle. Unscramble the corresponding letters to spell out the name of the item.

1. How many days in a week? _____

2. How many donuts in a "baker's dozen"? _____

3. How many sides to an octagon? _____

4. What day of the month is Valentine's Day? _____

5. What number is *seis* in Spanish?

   _____

PUZZLES AND PROBLEMS

35. Solid Characters

Animals represent the solid phase of matter. The names of one or two animals are hiding in each of the following words. See how many you can find.

a. board
b. scramble
c. steel
d. triumphant
e. billion

f. vacation
g. exasperate
h. whenever
i. femur
j. shareholder

36. Get a Clue

Identify each mystery element by combining the two clues.

First Clue	Second Clue	Mystery Element
three	electrons	1. _ _ _ _ _ _
eight	two	2. _ _ _ _
yellow	32.064	3. _ _ _ _ _ _
proton	one	4. _ _ _ _ _ _ _ _
little i	big S	5. _ _ _ _ _ _ _
balloons	noble	6. _ _ _ _ _ _

GA1691 Earth Science © Good Apple

PUZZLES AND PROBLEMS

37. Chemical Chaos

Part A: Use letters in the word *elements* to write the chemical symbols for three different elements. You may use each letter only once. As you identify each symbol, color in the corresponding letters; use a different color for the letters of each symbol. Then write the symbols and corresponding names in the spaces provided.

ELEMENTS

Symbol	Name
1. _____	_____
2. _____	_____
3. _____	_____

Part B: Which element has a greenish-yellow color, a strong odor, and is used to control the growth of algae in swimming pools? Use two of the remaining three letters from Part A along with six other letters to spell out the answer.

Mystery element: _____

PUZZLES AND PROBLEMS

38. Element Party

Combine the chemical symbols of two or more elements to spell out the answer for each clue. The first one is done for you.

a. Shorebird known as a sea swallow: <u>T</u> <u>e</u> <u>R</u> <u>n</u>

b. Word meaning "not cooked": __ __ __

c. Combination of two hydrogen atoms and one oxygen atom:

__ __ __ __ __ __

d. Frozen water: __ __ __

e. There are 206 of these in the human body: __ __ __ __ __

f. Nonflowering plant that reproduces with spores, not seeds:

__ __ __ __

g. Word meaning "not anybody": __ __ __ __ __ __

h. Word meaning "keen" or "invigorating": __ __ __ __ __

i. To sense sounds; to listen: __ __ __ __

j. Piece of clothing worn in cold weather: __ __ __ __

39. Phase Maze

An example of a liquid, solid, or gas is hiding in each of the following words. Underline the name of the matter, and identify whether it is a solid, liquid, or gas by circling *S*, *L*, or *G*.

a. biology	S L G		f. twice	S L G
b. neonate	S L G		g. centigrade	S L G
c. migraine	S L G		h. argonaut	S L G
d. repair	S L G		i. optimist	S L G
e. harmonica	S L G		j. harem	S L G

RHYMES AND RIDDLES

Part One: Rhyming Clues

40. Four-Letter Word
Thrust from the crust,
And needed for rust.

Answer: __ __ __ __

41. "Soot" Yourself
Number 6 on the Table,
With an organic label.

Answer: __ __ __ __ __ __

42. Sun Lover
Water, water, in crystalline form;
Won't last long 'cause it's getting warm.

Answer: __ __ __

43. "Sow" It Goes
The embryo grows in soil,
If nutrients within don't spoil.

Answer: __ __ __ __

44. Spread the Hue
A white light enters this silicated site,
And leaves in a spectrum of rainbow delight.

Answer: __ __ __ __ __

45. Heads or Tails?
A nickel here, a quarter there,
But save enough to pay the fare.

Answer: __ __ __ __ __

46. Pucker Up
It's shaped like an egg, juicy and sour;
It'll bring on a pucker with citric power.

Answer: __ __ __ __ __

47. Drop in Pressure
It falls from the sky and pelts the ground;
There's nary a dry spot anywhere in town.

Answer: __ __ __ __ __ __ __ __ __ __ __

RHYMES AND RIDDLES

48. Zoo Who?
Black and white, most bear-like too;
Keep the fish, toss me bamboo.

Answer: __ __ __ __ __ __   __ __ __ __ __ __ __ (two words)

49. Bit of Matter
A minus sign, yet out of sight;
Matter in orbit keeps things right.

Answer: __ __ __ __ __ __ __ __

Part Two: Ridiculous Riddles

50. What does a chemist who dabbles in real estate sell?

51. What part of a solid produces gas?

52. What holds most of the property of matter together?

53. What would you have if you spelled *vacuum* with only one *u*?

54. What makes up 88 percent of a particle?

55. What do crystallographers like in their salads?

56. What might occur if Fe combined with the element *Yttrium*?

57. What covers most of a solid?

58. How can you sell a solid?

59. What liquid is 75 percent of *soil* and produces gas?

60. What mineral is always part of a chemical change?

61. What part of *freeze* produces less stress on a car battery?

CHAPTER 2

Minerals

Minerals are inorganic substances that are the constituents of rocks. Minerals are pure and uniform, made of only one kind of material. They are classified by their chemical composition and the kind of chemical bonding that holds the atoms together.

Minerals

62. Dead Weight
Minerals are inorganic. They do not form from anything alive or once living. Use four letters in the word *inorganic* to spell the name of an important and common metal.

63. Nature's Way
Minerals are elements or compounds occurring naturally in the earth's crust. Where might a diamond be found in the phrase *occurring naturally*?

64. Tough Exterior
Minerals exist as solid forms. Which of the following minerals do you think is the most solid: talc, apatite, or diamond?

65. I Am What I Am
Which of the following chemical compositions cannot be considered a mineral? (Give a reason for your answer.) $CaCO_3$, $NaCL$, $BeAr$.

66. Internal Strength
Atoms that comprise a mineral always form the same pattern. For example, salt crystals are always shaped like a cube. What would you have if the internal pattern of atoms in the mineral *sodalite* reversed itself?

MIND STUMPERS

✦ ✦

67. Grand Tour
Tourmaline crystals often show striations, or lines. If all lines disappeared, the crystal would lose its identity. What would be left if one line was removed?

68. Down to Earth
Min E. Ralston, a seventh grader, collected rocks and minerals since she was in the fourth grade. She now has a fine collection of crystal specimens that include quartz, calcite, and halite. How many minerals appear in this paragraph?

69. Diamond Dilemma
Use a straight line to show a way in which a diamond can be cut into two equal halves.

Rule: You may not draw a picture of a diamond.

70. Let's Face It
Halite and galena form cubes. Show how both of these minerals can form the face of a crystal.

71. A Dash of Flash
Brian showed his teacher that he could make a three-pound (1.5-kilogram) chunk of hematite (iron ore) appear lighter than a six-ounce (170-gram) piece of gypsum. How did he do it?

THINK OF A WAY

72. HiYo Silver
Think of a way to show *silver* as a loose, thin strand by rearranging two letters.

73. Nickname
Tourmaline has ten letters in its name. Think of a way to describe this mineral using only three letters.

74. One Way
Cleavage is the way minerals split. Think of a way to show cleavage in one direction.

75. Artist at Work
Think of a way to draw a crystal with no sides.

76. One Here, One There
Think of a way to show galena (lead ore) composed of four parts lead.

77. No Fair
Think of a way to place the following words in the same sentence (use no more than six words): *quantity, quartz, quarry, quality.*

78. Still No Fair
Think of a way to place the following words in the same sentence (use no more than seven words): *mineral, malleable, mallet, mash.*

79. Totally Unfair
Think of a way to place the following words in the same sentence (use no more than six words): *cluster, cubical, crystals, colorless.*

80. Wait 'Til Spring
Think of a way to get calcite ($CaCO_3$) out of "a cocoon."

81. Tight Squeeze
Think of a way to show how four quartz crystals of various sizes can fill a 32-square grid.

THINK OF A WAY

82. Dirty Trick
Think of a way to turn tin into a louse egg.

83. Over Easy
Think of a way to turn metal ore into fish eggs.

84. Stake Your Claim
BAuxite is a mineral that is the principal source of aluminum.
Think of a way to make gold out of bauxite.

85. Think Fast
Think of a way to spell *silicon* by using the six-letter word *silver*.

86. Pinnacle of Superlative
A mineral collector looks for the perfect crystal. Think of a way to create
a "perfect" crystal by arranging seven different letters in a specific
pattern.

87. Iron Deficiency
Iron may form cubical crystals. Think of a way to form the face of a
crystal by using four irons.

88. Just Like Sardines
Native sulfur is soft, brittle, and fragile. Think of a way to fit 99 pieces
of S-shaped sulfur crystals into a one-centimeter square without the sulfur
pieces touching each other.

89. Look Hard
Think of a way to find more than one mineral in the word *mineral*.

90. Whose Got the Pull?
Copper wire is *ductile*—it can be pulled into thin strands without
breaking. Gold is an example of a ductile mineral. Think of a way to pull
nonmetal minerals, such as sulfur and asbestos, without breaking them.

91. Mineral Logic
Think of a way to create "fool's gold" from the following letters and
symbols:

e   ◈   t   π   r

PUZZLES AND PROBLEMS

92. Crystal Conference

Read the clues to help identify each mystery mineral. Rearrange the underlined letters to spell the name of the mineral. The first one is done for you.

a. Cub**i**c crys**t**als; sa**l**ty; **ea**sily crus**he**d. <u>HALITE</u>

b. **H**exagona**l** **c**rystals; **h**ardness of **t**hree; forms bo**i**ler s**c**ale.
   — — — — — — — —

c. **Ha**rdness: one; soaps**t**one; easi**l**y **c**ut.
   — — — — — —

d. Brown he**ma**tite; dull **l**uster; brow**n** or yel**l**ow; nev**e**r crys**t**all**i**zed.
   — — — — — — — — —

e. Var**i**ety of quar**t**z; dull, gray color; **In**dian arrowheads; used to **l**ight **f**ires. — — — — — —

f. **To**ugh, flexi**b**le, **el**a**st**ic; known as black m**i**ca.
   — — — — — — — —

g. Volcanic prod**u**ct; ye**l**low; **s**trong odor; **f**ound in fibro**u**s, ea**r**thy shapes.
   — — — — — —

h. Liquid bet**w**een z**e**ro and 100° cen**ti**g**r**ade; known as univers**a**l solvent. — — — — —

i. Small c**u**b**i**cal crys**t**als; red **c**opp**e**r or**e**; ha**r**dness: 3.5 to 4.
   — — — — — — — —

j. Iro**n** or**e**; magne**t**ic forc**e**; iso**m**et**r**ic cryst**a**ls.
   — — — — — — — — — —

93. Only Three

Unscramble the clues below to identify the names of three minerals found in the earth's crust. Confirm your answers by finding the mineral names among those listed in the box.

a. The name of a sticky mixture of water and earth, spelled backwards + "to move in a rapid fashion" + *co*.

b. The chemical symbol for sodium + "a strong wind."

c. H + the 15th letter of the alphabet + the 6th letter in *malachite* (copper ore).

copper	corundum	galena	gypsum
hypersthene	sodalite	water	calcite

PUZZLES AND PROBLEMS

❀ ❀

94. Two-Letter Getter

Identify each pair of missing letters by reading the corresponding clue. Rearrange the paired letters in each exercise to spell the name of a mystery mineral. The "big clue" offers additional information about the mineral.

Exercise A

1. A two-letter greeting: __ __
2. Fruit: __ __ ple
3. __ __ izzly bear
4. Wild duck: __ __ al

Big clue: pencil power
Mystery mineral: __ __ __ __ __ __ __

Exercise B

1. Van Gogh was an __ __ tist.
2. A girl's name: A __ __ e
3. A __ __ rcle is a geometric figure.
4. This crustacean has pincers: cr __ __

Big clue: number-one planet
Mystery mineral: __ __ __ __ __ __ __ __

Exercise C

1. The __ __ thon is a large, tropical snake.
2. Ductility is an example of __ __ nacity.
3. A city in France: Pa __ __ s

Big clue: "fool's gold"
Mystery mineral: __ __ __ __ __ __

Exercise D

1. To collect, gather: g __ __ an
2. Coming to an end: fi __ __ l
3. A large system of stars: __ __ laxy

Big clue: "Get the lead out!"
Mystery mineral: __ __ __ __ __ __

RHYMES AND RIDDLES

PART ONE: Mystery Verses and Challenges

Use clues in the verse to help you determine the name of the mystery mineral. Then see if you can solve the challenge problem.

95. Tough Customer

Mineral folks who bother to tell,
Say number 9 fits it so well;
As do crystals of hex persuasion.
It produces rubies to sell,
With sapphires "clear as a bell";
Welcome the emery abrasion.

Three additional clues:
1. The name rhymes with *bum, scum,* and *dumb.*
2. It produces six-sided (hexagonal) crystals.
3. It is number 9 on the scale of hardness; it is harder than any other mineral except diamond.

Mystery mineral: __ __ __ __ __ __ __ __ __

Challenge: The mystery mineral is comprised of a large amount of aluminum. Its chemical formula is Al_2O_3. What percentage of aluminum does the mineral contain? To find out, complete the following steps:

a. Look at a periodic table and find the atomic mass unit (emu) for aluminum and oxygen. Round off the values.
b. Add the total weights for two aluminum atoms and three oxygen atoms (Al_2O_3).
c. Divide the total weight from step *b* into the total weight of two aluminum atoms. This will give you the percentage of aluminum in the mystery mineral.

RHYMES AND RIDDLES

96. It's Everywhere

This mineral packs the crust,
From boulder to blinding dust;
Its name rhymes with bar and star.
As common as it may be.
Its hardness presents the key;
So think "felds" before the "par."

Mystery mineral: __ __ __ __ __ __ __ __

Challenge: The mystery mineral is the most common group of rock-forming minerals. *Microcline* is one member of the mineral group. Decode the following cryptic word to reveal the ten-letter name of another.

$$O + 18 + T + 8 + O + (6 - 3) + L + 1 + SE$$

PART TWO: Double Rhymers

Use the four-word clues to help you identify each mystery mineral. Read the "extra clue" for additional help.

97. Potluck: metal, sheen, kettle, green
Extra clue: It makes "cents."

Answer: __ __ __ __ __ __

98. Seeing Double: light, refraction, calcite, attraction
Extra clue: It's transparent and polarizes light.

Answer: __ __ __ __ __ __ __   __ __ __ __

99. Nature's Best: sapphire, ruby, admire, beauty
Extra clue: The mineral's name rhymes with *rum*.

Answer: __ __ __ __ __ __ __ __ __

100. Most Attractive: more, magnetic, ore, energetic
Extra clue: The mineral's name rhymes with *height*.

Answer: __ __ __ __ __ __ __ __ __

101. No Scent: sandy, poses, dandy, "roses"
Extra clue: It's a heavy mineral whose name rhymes with *kite, light,* and *sight*.

Answer: __ __ __ __ __ __

RHYMES AND RIDDLES

PART THREE: Ridiculous Riddles

102. What is the favorite mineral of boxers from the North Atlantic region?

103. Why is it hard to get lead out of galena (lead ore)?

104. Why can't gold ever be young?

105. In what form does change occur in sediment?

106. What mineral shows iron, but has no iron?

107. What mineral is 25 percent gas, but 100 percent solid?

108. How many minerals do you need to create a mine for a miner?

109. If diamond is harder than topaz, and topaz is harder than quartz, what is calcite?

110. *Geodes* are hollow rocks lined with crystals. What is in the middle of a geode?

111. What do mineralogists consider the most desirable part of asbestos?

112. What do you call a clock made only of copper?

113. What name was given to the female gem collector who labeled all of her mineral specimens backwards?

CHAPTER 3

Rocks

The earth's crust is composed of rock. Rocks are a combination of one or more minerals. *Igneous rocks* are made from hot liquid matter that has cooled and hardened. *Sedimentary rocks* are formed from accumulated layers of sedimentation in rivers, lakes, and oceans. *Metamorphic rocks* are formed when heat and pressure change one kind of rock into another.

MIND STUMPERS

114. Inside a Rock

The contents of a rock are scattered about. Rearrange the following word parts to write a three-word statement about what goes into the making of a rock.

Hint: The first word has seven letters, the second word has two, and the last has eight.

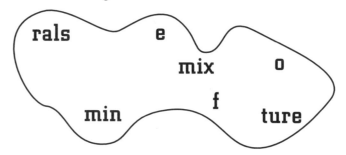

rals   e   mix   o   min   f   ture

115. Not Really

Look at a periodic table. If barium, sodium, and chlorine could combine, what rock might form?

116. New Identity

What planet would change to an "igneous rock formed deep in the earth" by adding *nic* to its name?

117. She Saw the "Life"

Susan had a lava rock with three cavities. Inside one cavity was a tiny zeolite crystal. Susan asked her brother Tom to examine the rock until he found it. A few minutes later, Tom yelled out, "I found it in all three cavities!" How could this be?

118. Tight Squeeze

Ron had five rock and mineral specimens: quartz, feldspar, mica, hornblende, and granite. He only had enough storage room for one specimen. Which one should he keep? Give a reason for your answer.

119. Model Example

A sedimentary rock is called a "classic rock" when it is a model of its kind; for example, a fine-grained red sandstone is a classic rock. *Clastics* are a group of sedimentary rocks composed of different-sized fragments. How can these rocks become classics?

CLASSIC ROCK

Rocks

MIND STUMPERS

120. Letter Go
Shale, under heat and pressure, may turn into slate. What two letters in *shale* undergo change before becoming *slate*?

121. Too Much Pressure
Joan Taylor introduced Dave Charles to Cybil Morphic. They shook hands. What just happened to Dave Charles?

122. Gather All Ye Pebbles
A *conglomerate* is a sedimentary rock composed of round pebbles cemented into a solid mass. Suppose you wanted to make a conglomerate rock for a science project. You gather several round pebbles and mix them with dirt and water. What would you be making as you stirred the ingredients together?

123. Rock Talk
Rock A asked Rock B, "Say, don't you hate the expression 'Dead as a rock'?" "Not really," answered Rock B. "You see, it wasn't always like that with me." What kind of rock is Rock B?

124. Weighted Down
Larissa rode to school on a unique vehicle. It had a seat of granite, a frame of basalt, and two tires of slate. Her sister, Madeleine, had a pet name for the vehicle. What do you think it was?

THINK OF A WAY

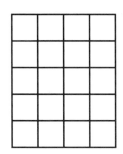

125. Asking a Lot
Think of a way to get the four-letter word *lava* to fill all 20 spaces in the grid. Each space in the grid can hold only one letter.

126. Ganging Up
Think of a way to show how heat and pressure might alter minerals.

127. Special Attraction
Think of a way to show the presence of limestone under certain conditions.

128. Take a Peak
Under heat and pressure, limestone becomes marble. Think of a way to show what marble looks like under pressure.

129. Tight Fit
Think of a way to fit *serpentine*, a non-banded metamorphic rock, into the following spaces:

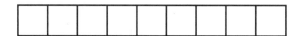

130. What's New?
The following picture shows the word *igneous* shaped like a hill or mountain. Think of a way to show a "new" picture by only removing four of the letters.

131. Who's to Blame?
Think of a way to show what might cause *limestone* to lose three letters.

132. X-Ray Eyes
Think of a way to see inside a geode without breaking it open.

133. In a Word
Concretions are round objects or odd-shaped masses of cemented material that collect around a nucleus. They are a feature of some sedimentary rocks. Think of a way to illustrate one by using the word *concretion*.

134. Particle Theory
The size of the particles in a rock determine the texture of a rock. Think of a way to show the difference between *fine texture* and *coarse texture*.

PUZZLES AND PROBLEMS

❀ ❀

135. Hide-a-Rock

Part A: Underline the two metamorphic rock names hiding in the following sentence: *Marsha Gagne is seven minutes late for her appointment.*

Part B: Combine two letters from each of the rock names to identify the missing word in each of the following sentences:

1. A __ __ __ __ is a biting insect with wings.

2. __ __ __ __ is fuzz made from the fine fibers of fabric or yarn.

3. To get the __ __ __ __ of an argument means to understand the main point or part.

136. Sentenced

Read the following sentences: *Sandstone is a sedimentary rock. It consists of cemented grains of quartz sand.*

Part A: Use parts of the words in the sentences to answer the following questions:

1. What is a form of precipitation? __ __ __ __

2. What is any branch of creative work? __ __ __

3. What is a vocal sound? __ __ __ __

4. What is another word for *against?* __ __ __

5. What is the name of a U.S. coin? __ __ __ __

Part B: Use the letters from your answers in Part A to write the name of an igneous rock.

Hint: It is a greenish-black or dark-gray rock with coarse grains and an even texture.

RHYMES AND RIDDLES

PART ONE: Mystery Verse and Challenge

137. Cork Power

A rock from an igneous vault,
A product of laval assault;
It shows off a silica coat.
In time, all flows come to a halt.
And really it's nobody's fault;
To witness a rock that can float.

Mystery rock: __ __ __ __ __ __

Challenge: Suppose the fine-grained, porous mystery rock absorbed a lot of water and froze. Why would only half the rock thaw, even if you put it in a 500° oven for 24 hours?

PART TWO: Triple Rhymers

Use the two sets of three rhyming clues to identify the name of each rock or event. If you still don't know the answer, read the "big clue" for additional help.

138. Tiny Grains
Big clue: smooth, slippery texture

| gray | clay | lay |
| silt | built | tilt |

Answer: __ __ __ __ __

139. Hot Spot
Big clue: a molten mass

| blow | flow | slow |
| vent | scent | spent |

Answer: __ __ __ __

140. Extrusive Event
Big clue: weapons and tools

| boom | doom | gloom |
| gas | mass | glass |

Answer: __ __ __ __ __ __ __

RHYMES AND RIDDLES

141. Fire Away

Big clue: hot liquid magma

diorite   syenite   pegmatite
conclusive   intrusive   extrusive

Answer: __ __ __ __ __ __ __

142. Over and Over

Big clue: long-term cycle or change

melting   cooling   grueling
change   exchange   rearrange

Answer: __ __ __ __   __ __ __ __ __ (two words)

PART THREE: Terse Verses

Use the clues in the title and the verse lines to help you determine the name of each rock.

143. Igneous Zone

If you really think I'm as "hard as a rock,"
Then set yourself for a bit of a shock.

Answer: __ __ __ __ __ __

144. Such Sediment

Half of my name is the same as a fruit,
But people who use me don't give a hoot.

Answer: __ __ __ __ __ __ __ __ __

145. Pressure on the Rocks

I'm formed under pressure, but must insist
Either "neis" or "nice," but never a schist.

Answer: __ __ __ __ __ __

146. Glacial Deposit

A mixture of boulders, sand, and clay,
Dropped by a glacier melting away.

Answer: __ __ __ __ __ __ __

147. Shale Away

First mud, then clay, plus nature's force;
And shale will change in time, of course.

Answer: __ __ __ __ __

RHYMES AND RIDDLES

Rocks

PART FOUR: Ridiculous Riddles

148. Why did the sandstone change to quartzite?

149. Ted held up four andesite rocks in his right hand and six rhyolite rocks in his left. His friend, Carol, said, "They all look alike to me. What's the difference?" What do you think Ted replied?

150. Which rodents lives in porous, glassy volcanic rock?

151. What did the psychologist say to the petrologist?

152. How do large masses of granite stay clean?

153. Where do many fossils come to rest?

154. Some land masses rise high above neighboring formations. What kind of experience is this?

155. Mr. and Mrs. Clay went to the Shales' house for dinner. They brought dessert. What kind of dessert did they bring?

156. The Clays and Shales went to a concert after dinner. What kind of music did they listen to?

157. Plants and animals buried in sediments may become fossilized. What do you call these organisms?

158. What limestone rock shows a sign of life?

159. When is quartz no longer quartz?

160. Why did John Pumice have trouble keeping friends?

161. Rocks appear in every part of the United States. In which state did igneous rock begin to form?

162. What path do people take as they travel through a cave?

CHAPTER 4

Volcanoes

A volcano is a mountain created by the flow of melted rock (magma) through an opening in the earth's surface. When volcanoes erupt, they release lava, hot gases, and dust. There are about 600 active volcanoes on Earth.

MIND STUMPERS

163. Keep It Together

Hot materials spew up through volcanic vents. As the hot substance builds up around the vents, various types of cones form. What do you think holds a cone together for so many years?

164. Hunt and Peck

Cinder cone, composite, and *shield* are three types of volcanoes. What do you think is a good way to type a volcano?

165. Ouch!

Volcano Bogus exploded. A huge ball of gas and ash weighing 923 pounds (419 kilograms) landed four miles away. What did it leave on the ground?

166. Super Magma

One magma flow produces no gas, but several magmas do. How is this possible?

167. Crustal Event

When many people hear the word *volcano,* they picture a volcanic cone with smoke swirling out of the crater. But it also refers to the actual opening in the crust where eruption occurs. Show how an eruption through the crust might look.

168. Clean and Clear

A volcanic eruption may produce lots of lava and shattered rock. How would an eruption need to occur in order to produce pure lava?

169. Lava Visit

One way lava reaches the surface is to move upward through an opening, or crack, in the earth's crust. Use the word *crack* to show how large and small rocks and gas move in an upward direction.

THINK OF A WAY

✿ ✿

170. Clone Dome

Think of a way to create a newly formed volcanic mountain from a second volcano.

171. Private I

Think of a way to reveal a volcano in *Hawaii* that's as old as the name itself.

172. Ash Dash

Black ash comes from some volcanic eruptions. Think of a way to create a sudden or violent reaction from black ash.

173. Give Mom a Break

Think of a way to show the presence of two "mothers" trapped in molten rock deep in the earth.

174. Something New

Think of a way to show how the word *volcano* written backwards produces salt.

175. Remove All Doubt

Think of a way to prove a "sure" fact about *fissure*—an open fracture around the base of a volcano.

176. Surprise!

When alternate layers of cinder and lava pile up, a composite cone develops. Scientists say a composite volcano is the most common volcano. Think of a way to prove that a composite volcano is really only half common.

177. Save the Neck

When an extinct volcano erodes, a volcanic neck of hardened magma may be all that remains. Think of a way to demonstrate this using the word *volcano*.

178. Share the Load

Think of a way to show how it takes two volcanoes to produce magma.

179. Cone Clone

All volcanic eruptions are not alike. Think of a way to show two volcanic eruptions that are alike.

180. Top of the Alphabet

A stiff, fast-cooling lava from Hawaiian volcanoes is known as "aa" (pronounced AH-ah). Think of a way to show how two Hawaiian volcanoes can produce "aa" lava.

Volcanoes

181. Bye, Bye, Magma
Complete the following explanation:

When magma erupts out of a volcano it becomes lava. Lava must do two things before it forms certain materials: first, it must lose heat; second, it must become rigid or solid. When lava loses heat, it __ __ __ __ __. Then it __ __ __ __ __ __ __ into a rigid or solid state to become __ __ __ __ __ __ __ __ __ __ __ __.

Hint: The first word rhymes with *spools*; the second with *gardens*. The letters needed to spell the first two words are located in the first and third rows in the following puzzle. To reveal the two-part name given to rigid lava, unscramble the letters in the second and fourth rows.

h	d	o	a	l	r
u	o	g	s	e	r
c	n	e	s	s	o
i	k	c	s	o	n

182. No Guarantee
The following terms indicate the "liveliness" of a volcano. Use the letters scattered within the illustrated volcanic cone to spell an adjective for each term. You may use letters more than once.

Term	Adjective
a. active	a. __ __ __ __ __
b. dormant	b. __ __ __ __ __ __ __ __

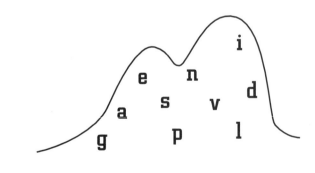

c. extinct	c. __ __ __ __ __

RHYMES AND RIDDLES

PART ONE: Mystery Verses and Challenges

Use the clues in the verse to determine the name of an active volcano.

183. Rumblings

Scientists say that subduction
Produced this monster eruption
of mudflows of ash mixed with snow.
A cascade site of destruction
Did much to limit production
When the "saint" decided to blow.

Name of the volcano:

__ __ __ __ __ __ __ __ __ __ __ __ __ __ __   (three words)

Challenge: The mystery volcano erupted in 1980. A bulge in the north side of the cone grew larger. When the bulge exploded, magma, steam, and gas escaped. What caused the bulge to explode?

To discover the answer, first identify the missing letters in the following puzzle. Then unscramble the letters to reveal the answer.

Idea: __ __ eory

Lock and __ __ y

Mineral: __ __ __ rtz

Muscle: h __ __ __ t

Answer: An __ __ __ __ __ __ __ __ __ __ caused the bulge to explode.

184. Quiet Time

A wide dome in tropical splendor
Once blew out smoke, ash, and cinder,
But now it's as dead as can be.
Rich soil grows onions most tender,
And fruit just right for the blender;
It's part of the Island Maui.

The extinct volcano: __ __ __ __ __ __ __ __ __ __

Challenge: The name of the extinct volcano means
"__ __ __ __ __ of the __ __ __." Interpret the cryptic picture to the right to discover the answer.

RHYMES AND RIDDLES

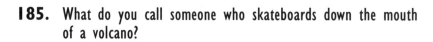

PART TWO: Ridiculous Riddles

185. What do you call someone who skateboards down the mouth of a volcano?

186. What do you call a volcanic opening that ceases to function?

187. A *caldera*—a volcanic crater—may develop when a volcano explodes and creates a basin-like depression. If hot steam burned someone during this event, what would be the result?

188. How many holes are there in a volcano?

189. What is the most disagreeable part of a volcano?

190. A laccolith can't become a batholith because a laccolith lacks something. What does a laccolith lack?

191. What would you get if you added an e to the opening from which lava erupts?

192. If you removed *cones* from *volcanoes*, what would be left?

193. What did the old volcanic cone say to the young volcanic cone?

194. A violent eruption can produce a *tsunami* (pronounced su NA mee)—a giant sea wave. If a tsunami washed over the Island of Rye, what would you have?

Volcanoes

CHAPTER 5

Earthquakes

Earthquakes are caused by the shift or movement of large rock masses, or *plates*, underneath the earth's surface. Most earthquakes occur along *fault lines*—fractures or breaks in the earth's crust. The magnitude of an earthquake (the amount of energy released) can be determined by using a Richter scale.

MIND STUMPERS

195. High Society

What part of society would be affected the most by a fault line running through the crust, as shown in the illustration?

C R U S T →

196. Gotcha!

Using a line or a circle, show how a seismograph is able to trap a layer of polluted air.

197. Center of Attraction

An *epicenter* is the point on the earth's surface where an earthquake begins. Show the presence of an earthquake in the middle of an epicenter.

198. Who's to Blame?

An earthquake occurred. Fault A said, "It's not my fault."
Fault B replied, "It's not mine either." Whose fault is it?

199. Short-Changed

The earth holds the source of earthquake energy. However, only 80 percent of the energy is available at any one time. Show how this is possible.

200. Not All There

Seismograms from three seismograph stations are needed to locate an epicenter. If only two stations report findings, what might be the results?

201. Animal Antics

Some animals act peculiar before an earthquake occurs, almost as if to warn others of the impending event. The following list of words describes some actions displayed by these animals. Find the hidden animal names in the descriptive words.

WORD	ANIMAL
a. irrational	__ __ __
b. defiant	__ __ __
c. unapproachable	__ __ __ __ __
d. unbearable	__ __ __ __
e. sheepish	__ __ __ __ __
f. embattled	__ __ __
g. flamboyant	__ __ __ __ or __ __ __
h. rambunctious	__ __ __

THINK OF A WAY

202. Falling Debris

Earthquakes can cause a lot of damage. Think of away to change the preceding sentence to describe earthquake damage to a plot of land.

203. Some Crust

Without drawing a picture, think of a way to show how a strong earthquake can change the crust.

204. Symbol Logic

Think of a way to use the following symbols to show how earthquakes may occur other than through movement along faults.

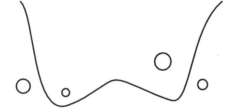

205. Crest Pest

Earthquakes in the sea floor may produce huge waves known as *tsunamis*. Think of a way to show how a tsunami becomes a great wave.

206. Not Again

Think of a way to use six words to show that earthquakes may occur frequently.

207. Big Shake

A severe earthquake may demolish a populated area. Think of a way to show how devastation can be great in size and extent.

208. Sick Leave

Think of a way to show how diseases may spread when a severe earthquake occurs.

209. Cliffhanger

A *scarp*—a small cliff—may develop during an earthquake. Think of a way to show how some healed skin tissue, a vehicle, and a freshwater fish contribute to the creation of a scarp.

210. Seismic Show

Think of a way to show how solar power makes up about half of a seismic sea wave.

211. In Shambles

Earthquakes can cause major damage to buildings, roads, and people. Earthquakes can be __ __ __ __ __ __ __ __ __ __. What is the most probable source of earthquake energy?

Use ten of the following scattered letters to complete the preceding sentence. Use the remaining four letters to answer the question.

s
i
a
d
o
u
t
s
e
a
h
r
s

212. Many Parts

Part A: Answer each question by using letters from the highlighted word. The first one is done for you.

1. What part of *fault* is oily or greasy? <u>FAT</u>

2. What part of *waves* is an abbreviation for a pathway or drive?
   __ __ __

3. What part of *epicenter* is a long narrative poem? __ __ __ __

4. What part of *magnitude* is the egg of a louse? __ __ __

5. What part of *intensity* is two less than a dozen? __ __ __

Part B: Use ten of the leftover letters from Part A to answer the following questions:

1. What word describes a shaking or trembling caused by an earthquake?
   __ __ __ __ __ r

2. What word describes the force behind an earthquake?
   __ __ __ r __ __

PUZZLES AND PROBLEMS

213. What Year?

Use the clues to identify each mystery word. Then find your answers in the puzzle and shade in the letter boxes. (Answers may be written up, down, horizontal, or backward.) The shaded area will reveal the year of the great San Francisco earthquake.

1. The name for the center of an earthquake: __ __ __ __ __

2. Part of the human head used for listening: __ __ __

3. Another name for *tear* or *cut*: __ __ __

4. An animal with a long nose: ant __ __ __ __ __

5. A tame or domesticated animal: __ __ __ __

6. The earth's closest star: __ __ __

7. An uninvited picnic guest: __ __ __

8. A children's game in which one player is "it": __ __ __

9. The abbreviation for an electroencephalogram: __ __ __

10. A vertical earthquake wave; a secondary wave: __ – __ __ __ __ __

11. Hot molten rock beneath the earth's surface: __ __ __ __ __

12. A large division of geologic time: __ __ __

13. Having proper respect for oneself: __ __ __ __ __ __

14. A flat drawing of the earth's surface: __ __ __

f	l	r	a	e	e	e	r	a	l	s	u	n
o	w	i	k	a	v	d	n	m	a	w	i	j
c	a	p	e	t	a	i	o	g	g	a	n	t
u	v	i	u	e	w	r	e	a	e	v	o	a
s	e	m	a	r	p	p	a	m	v	e	e	g

GA1691 Earth Science © Good Apple

RHYMES AND RIDDLES

PART ONE: Mystery Verses and Challenges

214. Record Keeping

Here comes a P-wave, an S- and an L-,
And, yes, they all have a story to tell;
The message comes straight from the crust.
Swirling dust and smoke leave a musty smell,
Damaged homes and roads don't show up too well;
The needle jabs data you trust.

The mystery verse describes a

— — — — — — — — — — — — —.

Challenge: Three horses—Parsley, Sagebrush, and Lilac—enter a race at Seismogram Meadows. Each of the three horses receives energy from an earthquake wave. One of the horses wins the race; the other two place second and third, respectively. Name the winning horse and the second- and third-place finishers.

Hint: Find out the speed of earthquake waves.

215. Get My Rift?

The crustal bed on the Pacific Coast
Is what this fault line craves the most;
Its relaxing sleep we adore.
With powerful reason to boast,
It turned San Francisco to toast;
Stay asleep, my friend (please don't snore).

The mystery verse describes the

— — —   — — — — — — — Fault. (two words)

Challenge: The mystery fault marks the boundary between two rigid, but moving, land plates. What are the names of the two plates?

Answers: 1. __ a __ i __ i __
2. __ o __ t __   __ m __ r __ c __ n (two words)

RHYMES AND RIDDLES

PART TWO: Double-up Rhymes

Identify the term that matches the description of the rhyming word pair.

216. Huge, deluge: __ __ __ __ __ __

217. Crack, wrack: __ __ __ __ __

218. Late, quake: __ __ __ __ __ __ __ __

219. Diastrophic, catastrophic: __ __ __ __ __ __ __ __ __ __

220. Force, source: __ __ __ __ __

PART THREE: Ridiculous Riddles

221. How is an earthquake similar to a young child eating dinner?

222. What do you call a mall for seismologists?

223. What is the difference between a P-wave and an S-wave?

224. Where do seismologists keep their valuables?

225. What is a seismologist's favorite music?

226. What two things produce an *earthquake?*

227. What is "foreshock"?

RHYMES AND RIDDLES

228. Why did Ima Fault leave home?

229. Why aren't there photos of earthquake waves?

230. What must an earthquake do before it is recognized?

231. How do seismologists greet one another?

232. What part of an earthquake does the most damage?

233. How would a baseball referee act during an earthquake?

234. How would a cow act during an earthquake?

235. How would a bird act during an earthquake?

236. How would a bee act during an earthquake?

237. How would a lion act during an earthquake?

CHAPTER 6

Forces Acting on the Earth's Crust

Natural and man-made forces acting on the earth's crust cause a breakdown, or decomposition, of rocks and minerals. Weathering—the physical or chemical process by which elements of the weather (e.g., rain, wind, frost) strip away layers of rock—is one of the major causes of change in the earth's surface.

MIND STUMPERS

❦❦❦❦❦❦❦❦❦❦❦❦❦❦❦❦❦❦❦❦❦❦❦❦❦❦❦❦❦❦❦❦❦❦❦❦

238. Mineral Magi

If chemical weathering occurs, a rock's minerals change into different substances. What would happen to the make-believe rock "dumicast" if chemical weathering caused it to lose all of its muscovite?

239. Full Force

A rock contains two minerals, x and y. It undergoes mechanical and chemical weathering. Using the symbols *xy* and *wz* to represent minerals, show how the weathering process affects the rock.

Hint: What's the difference between mechanical and chemical weathering?

240. Three Big W's

Stan W. Diver, geologist, described *erosion* as the breaking up and carrying away of material from the earth's crust. Three agents of erosion all have names that begin with the letter W. Write the name of each agent using letters from the geologist's name. You may use letters more than once.

241. From V to U

A glacier moving through a V-shaped valley can carve the valley into a U-shape. Use the word *glaCier* to demonstrate how this might look.

242. Ask Tom

The agents of erosion need something before eroding can begin. What do these agents need? If you ask Tom Oni for the answer, he will simply say, "Use the letters in my name."

MIND STUMPERS

❁ ❁

243. Water Power
Water moves weathered material from one area to another. Cross out the letters of the word *water* in the word *weathered*. Unscramble the remaining letters to identify the missing word in the following sentence:

"Take __ __ __ __, Marie. The river sediment is really dirty."

244. Gone Forever
A *glacier* is a body of ice that moves slowly down a valley or depression. If the large ice mass reaches the sea or lake, it breaks up to form icebergs. Write the word *glacier* on a sheet of paper, and draw a vertical line through the letter c to "break" the word into two equal halves. If you removed the letters that spell *large*, what would be left?

245. It's Gorge-ous
Erosion is the wearing away or removal of material by the action of water, glacial ice, or wind. What is the name given to a small, narrow ravine formed from erosion by running water? To find the answer, interpret the meaning of symbols A and B.

Hint: It's an ocean scene; interpret symbol A first.

246. Help Is on the Way
Alfred Wegener, scientist, believed that over 200 million years ago all the continents were joined together as a single landmass; over time, the continents drifted apart. He named the "world continent" Pangaea. Let's say that *Plateosaurus*, a long-necked dinosaur, lived on Pangaea. As Pangaea drifted apart and the landmass separated, *Plateosaurus* sent out a distress call in code. What was left of *Plateosaurus* after the call?

247. Spring Fling
Cheryl Rain and Jim Snow were crazy about each other. They were always together. What do you think they planned to do in the spring?

Hint: Think geology.

THINK OF A WAY

Forces Acting on the Earth's Crust

248. Falling to Pieces
Weathering is a natural breakdown and decomposition of rocks and minerals. Think of a way to show how granite may crumble into smaller pieces without undergoing any change in mineral makeup.

249. One at a Time
Weathering can lead to *exfoliation*—the peeling away of surface rock. Think of a way to use seven letters in *exfoliation* to spell a word that means "to separate into layers."

250. It's a Breeze
Wind scatters soil particles. These moving particles wear down solid surfaces with which they come into contact. Think of a way to describe wind in less than ten words.

251. Tons of Ice
A *glacier* is a moving body or huge block of ice crystals. Glaciers help erode the land by moving large amounts of rock and soil. Think of a way to form two blocks of ice crystals.

252. Ups and Downs
Diastrophism is the movement of the earth's crust. This movement may cause rock layers to fold, resembling the movement of ocean waves.

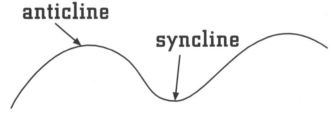

anticline

syncline

The crest (top) of the folding rock layer is called the *anticline*; the trough (bottom) is called the *syncline*. Think of a way to use the word *diastrophism* to show two anticlines and two synclines.

253. Table Talk
A *mesa* is a flat-topped hill with steep sides. A mesa may form when a plateau—a flat area higher than the surrounding land—erodes. Think of a way to rearrange the letters in *mesa* so that the word appears the same.

254. Slow Motion
In 1912, Alfred Wegener proposed a theory about how continents drifted apart and moved to where they are today. This idea became known as the Continental Drift theory. Think of a way to show continental drift using the word *continent*.

255. Ill Hill
Think of a way to illustrate an eroded hill without making a drawing.

PUZZLES AND PROBLEMS

✦ ✦

256. Pick a Part

Part A: Answer the following questions:

1. What part of *diastrophism* means "star" or "stars"?

   __ __ __ __ __

2. What part of *anticline* means "an involuntary muscular contraction"?

   __ __ __

3. What part of *plateau* means "water" in French? __ __ __

4. What part of *alluvial* means "by way of"? __ __ __

5. What part of *current* means "mongrel"? __ __ __

Part B: To identify the mystery word, fill in each blank with an unused letter from each of the highlighted words in Part A.

Hint: The mystery word is the name of a landmass that separated from Pangaea (the Continental Drift theory). This landmass was comprised of continents now know as South America, Africa, Australia, and India.

Mystery word: Go __ __ w __ __ __

257. Over Your Head
Using no more than twelve words, rewrite the following 35-word paragraph so that it makes more sense.

In many caverns, _____ (ground/lakes) are formed

as ground _____ (two hydrogen atoms and one

oxygen atom) moves through _____ (small, lemon-

shaped, greenish-yellow citrus fruit + hard, solid, nonmetallic mineral

matter of which rock is composed).

RHYMES AND RIDDLES

PART ONE: Mystery Verse and Challenge

258. Most Attractive

Look at the boulders at the base of the hill.
It's talus, of course; don't mistake it for till.
What force do you think pulled them down?
It's the same force that helps water spill,
And permits awesome landslides to kill.
We're held to the ground most earthbound.

Mystery force: __ __ __ __ __ __ __

Challenge: Let's add some spice to the mystery force by changing its name to the word *variety*. What must be done to make this happen?

PART TWO: Ridiculous Riddles

259. What can pull objects without any hands?

260. What happens to some rock layers as stress builds?

261. What is the most stimulating part of the Plate Tectonics theory?

262. What helps some mountains stay cool during the summer?

263. Why is the wind considered very particular?

264. What legless form of matter can run at any speed?

265. Why is it important to have part of a particle?

266. Where is the wettest part of a melting glacier?

267. Barbara Boulder and Rhonda Rock go to the playground almost every day. What is their favorite activity?

CHAPTER 7

Prehistoric Life on Earth

The earth is about 4.6 billion years old, but scientists conclude there was no life on this planet until around 3.5 billion years ago. Fossils found in rocks help paleontologists determine the age of different prehistoric plants and animals. The oldest animal fossils are that of crustaceans, then fish, followed by amphibians, reptiles, and mammals. Most of the ancient plant and animal species are now extinct, including the dinosaurs—huge prehistoric reptiles that ruled the earth over 200 million years ago.

MIND STUMPERS

268. Pinpoint
Use four different letters in the word *fossil* to identify a particular kind of earth.

269. Name Game
Paleontologists are scientists who study prehistoric life and fossil remains. Where would a paleontologist display a period of time that is greater than an era?

270. One Way or Another
Show four ways an animal might look if only one-half of it became a fossil.

271. Tough Call
Usually only the hard parts of a plant or an animal become fossils. What is the hardest part of a *Cactocrinus nodo brachiatus* (sea lily)?

272. It Makes Sense
A fossil of *Archeopteryx*, a prehistoric bird from the Jurassic period, showed the presence of wings and feathers. Yet after examining the fossil, some scientists believed it probably couldn't fly. Why would they draw that conclusion?

273. Big Surprise
Paleontologists try to reconstruct a picture of how extinct plants and animals lived. In all cases, what did they discover about a fossil animal's eating habit?

274. Fond of Fauna
Many of today's animals are protected in a game preserve. Flora Fauna, a scientist, was in charge of a special building designed to protect animal fossils. How did she refer to the dwelling?

Hint: Look at the last two words in the first sentence.

275. Barely There
How much of a fossil would you need to recover to recognize your find as a fossil?

276. A New View
Sean spent five years collecting sea-life fossils. His friend, Marlene, asked him how he grouped his specimens. Sean smiled, picked up a pencil, and scratched the following markings on a piece of paper. What was Sean trying to tell Marlene?

sǝʇɐɹqǝ

277. Take a Bite
What extinct marine arthropod is related to the crab? Use the clue inside the box below.

lobite^3

THINK OF A WAY

❧ ❧

278. Follow the Pattern
Think of a way to show a fossil formation with four fossils.

279. Falling Apart
Think of a way to illustrate a poorly preserved fossil using the following symbols:

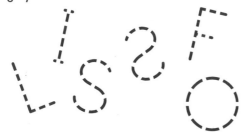

280. You Never Know
Only once-living organisms can become fossils. Therefore, an iron nail doesn't meet the requirement. Think of a way a nail might end up as a fossil.

281. Roach Approach
Cockroaches swarmed during the Pennsylvanian period (part of the Paleozoic era). Overcrowded conditions must have led to a dull life packed with pain. Think of a way to indicate distress on cockroaches.

282. Upward Mobility
During the Cenozoic era, mammals rose from obscurity to the dominant position in the animal kingdom. Think of a way to show how this might appear.

283. That's It, Period.
A *period* is a subdivision of a geological era of Earth history. For example, the Paleozoic era features seven periods of time. Think of a way to show the number of periods in the Mesozoic era without writing words or using numbers.

284. Too Many Letters
Think of a way to describe a dinosaur by using one *p*, one *b*, two *t*'s, two *l*'s, two *i*'s, three *r*'s, and four *e*'s.

Hint: What does the word *dinosaur* mean?

285. A Star Is Born
Think of a way to show how starfishes originated early in the Paleozoic era.

286. A Bit of Wizardry
Think of a way to reveal a mythical land nestled in the middle of two major divisions of geological time.

Hint: Era.

PUZZLES AND PROBLEMS

❧❧❧❧❧❧❧❧❧❧❧❧❧❧❧❧❧❧❧❧❧❧❧❧❧❧❧❧❧❧❧❧❧❧❧❧❧❧❧

287. Reveal Yourself

A *fossil* is preserved evidence of past life. Put the following clues together to identify two organisms whose "remains" are preserved in the illustrated rock layers.

Hint: Combine the phrases and letters to spell the names of the two organisms.

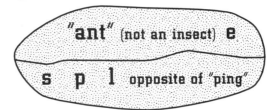

"ant" (not an insect) e

s   p   l   opposite of "ping"

288. Modern Times

Part A: The Cenozoic era is known as the "Age of Mammals." Find the names of eight different mammals by reading the clues.

1. Stripes, hooves: __ __ __ __ __

2. Mane, king: __ __ __ __

3. Fur, hibernate: __ __ __ __

4. Race, gelding: __ __ __ __ __

5. Horn, huge: __ __ __ __ __

6. Hump, desert: __ __ __ __ __

7. Bovine, buffalo: __ __ __ __ __

8. Vertebrate, *Homo sapien*: __ __ __

Part B: Use the third letter in each of the first four answers from Part A to spell the name of a wild mammal that lived in the Cenozoic era.

Answer: __ __ __ __

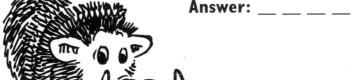

RHYMES AND RIDDLES

PART ONE: Mystery Verse and Challenges

289. Frost Bite

*My rise to distinction
Is based on extinction;
Ice crystals cover the ground.
I'm bulky and hairy,
With curved tusks most scary;
So awesome . . . pound for pound.*

Mystery beast:

— — — — — — — — — — — — — — — (two words)

Challenge 1: During the Pleistocene epoch of geologic time, the mystery animal lived on frozen land coated with ice. Count how many times you can produce *ice* from the letters in the following puzzle. You may use each letter only once.

Hint: Look for various forms of precipitation. Remember, ice occurs in various forms.

w	c	i	l	e	s	e	a	i	t	c	e	i	e	i	p
e	i	c	h	m	z	r	f	e	e	z	o	t	e	c	c
z	c	i	e	j	o	e	i	c	a	r	e	c	e	b	r
i	s	e	i	j	r	n	k	k	f	c	p	w	a	i	n

Challenge 2: Find the flying-insect "mother" hiding in the mystery animal's name.

Challenge 3: An elephant is a cousin of the mystery beast. Think of a way to use 20 letters to show the results of crossing an elephant with the mystery animal.

RHYMES AND RIDDLES

PART TWO: Ridiculous Riddles

290. Why did the paleontologist have a hard time finding a job?

291. What did the mold say to the cast?

292. What extinct marine organism is nearly half bite with no teeth?

293. How do trees feel about becoming fossils?

294. What did the 30-million-year-old fossil say to the 140-million-year-old fossil?

295. How did the celebrity fly encased in ancient tree resin sign her autograph?

296. Why do paleontologists who do impersonations flop at parties?

297. *Scanty* means "meager" or "not enough." What animal always appears in a scanty fossil find?

298. Why are cephalopods (octopus and squid) more apt to become "perfect" fossils?

299. Why did giant dragonflies living in the Paleozoic era walk rather than fly?

300. The first birds appeared during the Jurassic period (part of the Mesozoic era). What came after them?

301. What has no tongue, no lips, and no larynx, but can tell us about past life?

CHAPTER 8

Weather and Climate

Weather is the day-to-day variation of climatic conditions at a particular place. Factors that determine weather conditions include atmospheric pressure, temperature, humidity, wind, cloud cover, and precipitation. Weather forecasts are based on data collected from weather satellites and compiled by computers at weather agencies around the world.

MIND STUMPERS

302. Grab Your Hat

Sometimes strong vertical winds called *downdrafts* occur during thunderstorms. Show how this might look over a mountain range.

303. Miracles Never Cease

It has been said that nothing's impossible. If so, how might someone duplicate a cyclone?

304. Foil a Funnel

Use three X marks to show how to prevent a tornado from forming on a prairie.

305. No Umbrella Needed

How can an amateur rainmaker create rain without any experience or equipment?

306. Turn on the Fan

Rearrange the letters in *weather* to show a warming trend.

307. She's "Bean" Around

Ms. Wong asked her students what they knew about climates of the world. Corina said, "Half of the climates anywhere on Earth are leguminous." What did she mean?

308. Dreary Days Ahead

In the make-believe town of Clement, Ohio, Nancy Cloud reports the weather every morning on radio station KWIG. She always closes her report with ". . . Tune in tomorrow for more in Clement weather coming your way!" A recent survey showed a 60 percent drop in Nancy's listening audience. What could be the reason?

309. The Weather Store

How many words pertaining to weather and climate can you form with the letters in *troposphere* (the lowest layer of the atmosphere). You may use each letter only once.

310. Hidden Agenda

In the following list of letters is a hidden four-word sentence that describes a certain type of cloud. Circle the letters that spell the words of the descriptive sentence.

Hint: The sentence consists of a two-part noun, a verb, and a "fluffy" adjective.

ciutmeuvleusactlmoiuxdirsrawrecpeulfvfuy

THINK OF A WAY

❧❧❧

311. Bring It Together
Think of a way to unscramble these sentences into two short poems. One describes a hurricane; the other, a tornado.

It can pick up a tree and farm. And offers a "big eye" of charm. It sets off a major alarm. And will do it with no hand or arm.

312. Stretch It Out
The *climate* is the average weather condition at a particular area over a long period of time. Think of a way to illustrate *climates* with six letters.

313. Birth of a Breeze
Think of a way to depict the origin of winds.

314. Self-Destruct
Think of a way to describe damage done in the word *tornado* itself.

315. Can't Miss
Think of a way to show how weather changes on a daily basis.

316. Equal Shares
Think of a way to divide the atmosphere into four layers.

317. Mist Again
Water exists in the atmosphere as a solid, a liquid, and a gas. Think of a way to show water changing from one state to another.

318. Economize
Humidity is moisture in the air. Think of a way to cut humidity by one-fourth.

319. Come on Out
Think of a way to show precipitation within *precipitation*.

Weather and Climate

PUZZLES AND PROBLEMS

❧❧❧

320. The Right Find

Read each of the following clues to help you determine the missing rhyming word. The first one is done for you.

a. Looking up into the stratosphere.

sky,   <u>high</u>

b. The meteorologist goofed—the predicted stormy weather gave way to fair skies.

thunder,   b __ __ __ __ __ __

c. Liquid water to water vapor; water vapor to cloud droplets.

evaporation, c __ __ __ __ __ __ __ __ __ __ __ __

d. Air cools below the dew point and combines with smoke and other pollutants.

fog,   __ __ __ __

e. Whirling air and debris moving across the desert.

dust,   __ __ __ __

f. A thunderstorm arrives with a strong wind.

hail,   __ __ __ __

321. A Time to Get Wet

Part A: *Nimbus* refers to rain clouds. How many spellings of the word *nimbus* can you find in the following list of letters?

n, b, u, n, s, r, m, i, b, i, n, s, a, i, m, u, h, c, m, u

Part B: Unscramble the remaining letters from Part A to reveal a dreary weather forecast.

Weather forecast: __ __ __ __ __ __ __ __

322. Hint Parade

Use the clues to help identify three weather instruments.

a. A measurement equivalent to 39.37 inches + heat:

__ __ __ __ __ __ __ __ __ __ __

b. An apparatus for measuring + the chemical symbol Hg + rise + fall:

__ __ __ __ __ __ __ __ __

c. "Mother" + air in motion + gauge:

__ __ __ __ __ __ __ __

RHYMES AND RIDDLES

❧ ❧

PART ONE: Mystery Verses and Challenges

323. Too Humid

Those "down in the dumps" know my name,
Warm, moist air with plenty of rain;
A low-pressure belt, to be sure.
I bring thunderstorms over Congo game,
And drench dense forests of Amazon fame;
I'm something without an allure.

The mystery verse describes the __ __ __ __ __ __ __ __.

Challenge 1: Complete the following sentence by unscrambling the last six words.

The mystery place identified from the verse is found near the geographic area described as "eth eratg irlecc fo het trahe."

Challenge 2: From left to right, circle the letters that spell out the geographic location where one can find the mystery place.

a c e f m q i l u d x a p u t r e o r

324. Polar Power

Jack Frost just loves it up here,
With ice fields of snow and reindeer;
Mosses and small shrubs abound.
Dark winter creeps through the year,
And crowds out the summer so dear;
There's nary a suntan around.

Mystery polar region: __ __ __ __ __ __

Challenge: In the mystery polar region, less than _____ inches of rain fall during the year. To find the amount, use the clues from the following rhyme:

Use atomic number fifty,
And find the answer quickly.

RHYMES AND RIDDLES

PART TWO: Ridiculous Riddles

325. How do people who enjoy sailing refer to wind?

326. Why can't the layers of atmosphere be somewhere else?

327. Why is wind speed on a weather map so gnarly?

328. What is feathery and stays high in the sky, but cannot fly?

329. What do evaporation, condensation, and precipitation have in common?

330. Why wasn't John Wilson, meteorologist, an easy person to get to know?

331. Why are weather forecasters never wrong?

332. Why did co-workers stay away from the weather forecaster?

333. What kind of barometer measures air pressure on the first planet from the sun?

334. How are thunderstorms like small children?

Weather and Climate

CHAPTER 9

Astronomy

Astronomy is the study of our universe. Astronomers research the many components of space, including stars, asteroids, meteors, comets, and the nine planets that make up our solar system: Mercury, Venus, Earth, Mars, Jupiter, Saturn, Uranus, Neptune, and Pluto.

MIND STUMPERS

335. Moon Bloom
If someone asked you to draw the outline of a doughnut, the sun, or the moon, you'd probably make a circle. How would you show a full moon without making a sketch or drawing?

336. Not Easy to See
Venus, the second planet from the sun, is immersed in an atmosphere of carbon dioxide. How can you show the presence of at least one volcano on the surface of Venus?

337. Astronomical Task
Suppose you've been asked to reveal a hole in the middle of Jupiter. Unfortunately, you must complete the job without a telescope, photographs, or any space equipment. You may only use paper and pencil. How would you do it?

338. Planet Parade
How many planets can you spell with the following letters? You may use letters more than once.

Clue: There are more than five names to spell.

p m a r s s u n e v n e t a l

339. Meeting Place
Without drawing an illustration, show four different meteors crossing a comet's path.

340. Wrong Way
Use your creativity to show how a comet moves backward through the sky.

MIND STUMPERS

✦ ✦

341. Smaller and Smaller

In the make-believe town of Planetville, astronomer Dr. Sara Solar discovered a large planet that contained four smaller planets. Each of the smaller planets had a different diameter. Rather than reveal the name of her new find, Dr. Solar chose to keep it a mystery. However, she *did* identify the six letters needed to spell the name of the planet.

A   G   T   E   T   R

Make a sketch of the mystery planet from the information stated about its size. If you do so correctly, you will learn the name of Dr. Solar's planet.

342. Another Surprise

Dr. Tom Nova, astronomer, discovered a new planet eight years after Dr. Solar's find. He named it "Dyrtytub." Six days later he made another startling discovery about the planet. What do you think it was?

Hint: The unscrambled name reveals a clue about the discovery.

343. Boxed Planets

Uranus, Neptune, and Mercury share something in common. What is it?

Hint 1: Silicon and metal combine to produce the "boxed in" secret.
Hint 2: You will never see the three planets without it.

344. Above the Belt

Use letters from the phrase *Planet Earth, Year Two Thousand* to write the names of at least seven human body features located from the waist up. You may use letters more than once.

345. Meteor Invasion

How many spellings of *meteor* can you find among the scattered letters in the following figure? Before you determine the answer, divide the figure in half by drawing a line between the two X marks.

Hint: Observe what you get by separating the two halves.

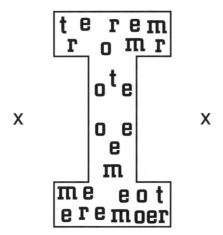

THINK OF A WAY

346. The Break of Day
Think of a way to show the night sky in the middle of the day.

347. It's Not a Belt
Think of a way to show what holds the e's on an eclipse.

348. A Cheese Ball?
Think of a way to expose an unidentified flying object in a full moon.

349. Time's Up
Think of a way to reveal the earth's closest star in two different planets.

350. Of This Earth
Think of a way to show how lack of movement keeps all terrestrials together.

351. Day Thief
There are 365 days in a year. Think of a way to show a year with only 273 days.

352. Never the Same
Stars that show a large change in brightness are called *variable stars*. Without drawing a picture, think of a way to show four ways in which a star demonstrates variety.

353. Now We Know
Craters exist on the surface of the moon. Think of a way to reveal what rests in the middle of all craters.

354. Female Power
A strong woman can do almost anything. Think of a way to show how female strength holds a constellation together.

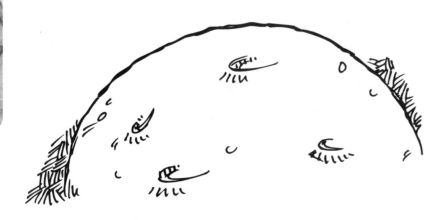

PUZZLES AND PROBLEMS

❀ ❀

355. Constellation Presentation
Combine the letters, symbols, and words in each of the following problems to identify the name of a constellation. For an extra clue, read the nickname in parenthesis.

1. + o + a golfer's score + *dalis*

   Answer: __ __ __ __ __ __ __ __ __ __ __
   (the Giraffe)

2. A "precious stone" + opposite of *out* + the 9th letter of the alphabet

   Answer: __ __ __ __ __ __ (the Twins)

3. The chemical symbol for gold + the first two letters of a word that is the opposite of left + the chemical symbol for the element with atomic number 31

   Answer: __ __ __ __ __ __ (the Charioteer)

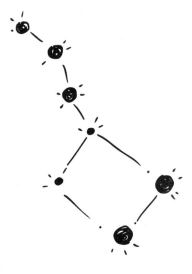

356. Where's an Astronaut?
Unscramble the letters in each exercise to spell a term known to astronomers. Use the five-word clues to help you identify the terms.

1. **Scrambled word:** jaonvi
   Clues: planets, Jupiter, Saturn, Neptune, Uranus

   Answer: __ __ __ __ __ __

2. **Scrambled word:** lenapohi
   Clues: point, Earth, sun, farthest, orbit

   Answer: __ __ __ __ __ __ __ __

3. **Scrambled word:** dzcaio
   Clues: constellations, Leo, Virgo, belt, Pisces

   Answer: __ __ __ __ __ __

4. **Scrambled word:** eblaun
   Clues: sky, dust, gas, space, outer

   Answer: __ __ __ __ __ __

5. **Scrambled word:** nadpoteil
   Clues: Jupiter, Mars, belt, asteroid, telescope

   Answer: __ __ __ __ __ __ __ __ __

RHYMES AND RIDDLES

PART ONE: Mystery Verses and Challenges

357. One of Nine

I linger 'tween six and eight,
With narrow rings part of my fate;
A blue-green cloud is all I see.
Herschel caught me in his lens,
But missed my "mooniest" friends;
The earth is less massive than me.

Mystery planet: __ __ __ __ __ __

Challenge 1: The mystery planet was discovered by Herschel in 1781. To reveal Herschel's first name, put together the answers to the following clues:

1. An upside-down *M*.
2. A three-letter word for *sick*.
3. Replace the letter *H* in *ham* with the letter *I*.

Answer: __ __ __ __ __ __ __

Challenge 2: The mantle of the mystery planet is made of rock, water, methane (CH_4), and ammonia (NH_3). The atmosphere surrounding the mystery planet consists mainly of hydrogen and helium. In the following equation, use the atomic mass for carbon, nitrogen, hydrogen, and helium to determine how many Earth years are equivalent to one year on the mystery plant. (Round off the mass values to the nearest whole number.)

3(atomic mass of nitrogen) + 2(atomic mass of carbon) + 3(atomic mass of helium) + 6(atomic mass of hydrogen)

RHYMES AND RIDDLES

358. Some Galaxy

Billions of stars in motion,
Like sand grains in the ocean;
Puncture the sky with ease.
With plenty of dust and gas,
And planets of sizable mass;
Somewhere you'll find Pleiades.

Name of the well-known galaxy:

__ __ __ __ __   __ __ __ (two words)

Problem: What kind of structure or shape does the mystery galaxy have? Correctly unscramble the following letters to reveal the answer.

Hint: It curves.

```
        P
   L        I
S    A    R
```

Answer: __ __ __ __ __ __ __

359. Space Debris

We travel among the stars,
Mostly 'tween Jupiter and Mars;
Our presence will surely be felt.
Ceres and Vesta at large,
Together we lead the charge;
And weave a heavenly belt.

Mystery objects: __ __ __ __ __ __ __ __ __

Problem: The sizes of the mystery space debris range in diameter from less than one mile to over **600** miles. About how many of these objects do you think exist in our solar system? To find out, solve the following problem:

$13 + 100 + 144/12 + 75 + (4 \times 2 \times 10^2)$

RHYMES AND RIDDLES

PART TWO: Ridiculous Riddles

360. Joan has two terrestrials. Bill gives Joan one of his terrestrials. What does Joan now have?

361. Why do all stony meteorites weigh 2,000 pounds (907 kilograms)?

362. How do astronauts keep their pants from slipping?

363. What is the difference between the "morning star" and the "evening star"?

364. Why does the moon act strange at times?

365. How can a meteor add three letters to its name?

366. How many "full moons" in one *MOON?*

367. What moon phase emits a distinct fragrance?

368. Where can a canine be found on Mercury?

369. What would you have if an eclipse lost its lip?

CHAPTER 10

Oceanography

The Pacific, Atlantic, Indian, and Arctic Oceans cover approximately 70 percent of the earth's surface. The salty waters range in depth from about 600 feet (180 meters) to over 36,000 feet (10,800 meters). A wide variety of plants and animals live in the ocean, including mollusks, arthropods, fish, amphibians, reptiles, and mammals.

MIND STUMPERS

370. What, No Pepper?

If you removed NaCL from salinity, what would remain?

371. The Two W's

Winds and waves churn ocean water near the surface. This is called mixed _____. Use the illustration to help you find the answer.

$$Y \quad E \quad L \quad \begin{array}{c}R\\A\end{array}$$

372. Ride the Tide

Marcella gave an oral report to her science class about the discovery of a bottle washed ashore. The message in the bottle indicated that the note was written 20 years ago. She jokingly described her report as a "current event." What did she mean?

373. Chemistry Humor

Diatoms are one-celled water plants that have cell walls of silica. Charles Wilkins, chemistry teacher, referred to these plants as "a thingamajig having two atoms." What did he mean?

374. Absurd Bird

Carl and Sue went to lunch at Mario's Deli. Carl ordered cold, oily red meat on rye and no tomato. His meal arrived ten minutes later. Carl was surprised to find a mischievous cormorant (a diving water bird) hiding in his order. Silly, isn't it? How could this possibly happen?

375. Turn on the Light

The fast-moving scallop can spot danger with its 30 or 40 eyes. Find a way to show a scallop that has no eyes, yet is able to see.

376. Over Your Head

Hidden within the following sentence is a word that describes an area of the ocean depths where seamounts and guyots are found. To reveal the name, unscramble the five underlined letters.

*Some **a**reas of the ocean **b**ottom support **s**eamounts—submerged mountain peaks; the ones with flat-topped peak**s** are called gu**y**ots.*

377. What Next?

Atoll refers to a circular coral reef that completely (or almost completely) surrounds a lagoon. Suppose there was a bridge leading to the reef-encircled lagoon. What would be ironic about charging people a fee to cross the bridge?

THINK OF A WAY

378. Higher and Lower

Salinity refers to the amount of salt and dissolved minerals in water. Think of a way to show how the salinity level of water can be lowered and raised at the same time.

379. Adore the Floor

The term *benthos* refers to plants and animals living on the ocean floor. Think of a way to show life on the ocean floor.

380. Floaters Anonymous

Many ocean animals eat *plankton*—tiny floating organisms. What part of a plankton allows it to drift with the ocean current?

381. Snuggle Up

Think of a way to prove that mussels and barnacles display emotion.

382. Seaweed Be Gone!

Algae has clogged the mouth of Winston Bay. Someone needs to thin it out. How about giving it a try? Think of way to reduce the amount of algae without using any material other than a pencil and paper.

383. Free Swimmers

Think of a way to use the letters in *wish, leaf,* and *heel* to write the names of three *nekton*—free-swimming aquatic animals.

Rule: You must use all of the letters.

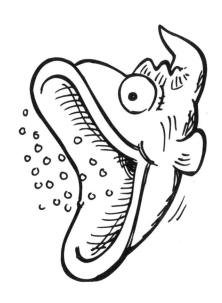

PUZZLES AND PROBLEMS

384. Grasp Task

Use all the following letters to spell the name of a long, slender, flexible growth on the head of a squid.

Hint: Look for a five-letter pattern.

e	a	t	a	c	t	l	t	c	t
c	e	t	e	l	a	c	a	l	e
t	a	e	c	l	t	l	c	e	t
c	t	l	a	e	e	c	l	l	c
a	l	l	a	a	t	a	e	c	e

385. Whale Tale

See how many times you can spell *orca* ("killer whale") by using all of the following letters:

```
      r r
   r a o r a
 c r o c c o c
   a c a a
   o o a c r
```

386. Pelagic Magic

Pelagic is a term that describes anything related to or living in the marine area just above the ocean floor. Use the letters in *pelagic* along with the following rhyming clues to spell the names of various organisms that live in the pelagic zone of the ocean.

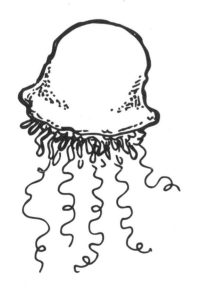

Portuguese man-of-war

1. It rhymes with *meal.* __ E __

2. It rhymes with *stale.* __ __ __ __ L __

3. It rhymes with *nail.* __ __ A __ __ __

4. It rhymes with *sea.* __ __ G __ __ __

5. It rhymes with *bearing.* __ __ __ __ I __ __

6. It rhymes with *applaud.* C __ __ __

PUZZLES AND PROBLEMS

387. Diving into Cormorants

A *cormorant* is a large diving bird with webbed toes and a hooked beak. Japanese fishermen use cormorants to help them catch fish. The fisherman ties a string around the bird's foot to keep the animal from flying away with the catch.

Problem: Somebody locked a cormorant in a box. It's up to you to free the bird. First, use the letters scattered in the box to spell the bird's name. Next, arrange the remaining four letters to spell the name of a body structure designed to allow the bird to fly away. Last, answer the questions. (Be on your toes: A trick "flutters" overhead.)

```
c i a t g r w
o n m r n o
```

a. What do the remaining letters spell? ___ ___ ___ ___
b. Were you able to free the bird by spelling its name?
c. Were you able to help the bird fly away? If not, why?

388. A Closer Look at Cormorants

In front of each body-part name, write the letter of the matching description. Then complete the picture by drawing a cormorant standing on the rock.

Body Parts	Descriptions
____ 1. bill	a. long
____ 2. foot	b. wedge-shaped
____ 3. color	c. hooked
____ 4. neck	d. webbed
____ 5. body	e. slender
____ 6. tail	f. dark

RHYMES AND RIDDLES

PART ONE: Mystery Verse and Challenge

389. Floating Along

*I cherish the ride
in low and high tide,
Yet a fish I cannot be.
With no fins for motion,
I drift in the ocean,
So silent and gracefully.*

Mystery organism: __ __ __ __ __ __ __ __ __

Challenge: The free-floating mystery organism is also called a *medusa*. A medusa travels great distances as it drifts with the ocean current. Which country supports the medusa no matter where it drifts around the world? (Be aware: A trick is in the air.)

PART TWO: Ridiculous Riddles

390. Which tentacled sea animal has its top in the middle?

391. Which shelled organism is always alone, no matter where it goes?

392. Which reef structure needs coal to hold it together?

393. I can rise or fall . . . or barely move at all. What am I?

394. In what condition would you find a long rolling wave on the open sea?

395. A commercial fisherman hauled up 45 sole in his net. One of the fish flopped off the deck and fell back into the water. What would you call the lucky fish?

396. Which undersea structure amounts to more than any other ocean feature?

397. What did the starfish say to the jellyfish?

398. What do you call a person who uses a telescope and oxygen tank at the same time?

399. When do ocean waves experience depression?

400. What is the favorite billiard game of marine biologists?

401. Which sea animal has a distinct chin, but no face or head?

CHAPTER 1: MATTER

1. Use letters from the phrase *matter so* to spell the word *atoms.*

2. It shows water going from one phase to another.

3. Spring.

4. There are two: one C (carbon) and one O (oxygen).

5. There are 14 e's in the 111 chemical symbols. However, there are only four e's in the phrase *the elements.*

6. *Sweat + er = sweater.*

7. Use a pencil to lightly sketch a side view of an evaporating dish. Then slowly erase the drawing. Thus, the dish is evaporating *(groan).*

8.

9. Subtract 10.2 from 207.2 to get 197.0, the atomic mass for gold.

10. Change the letter *m* to the letter *f* to spell *fountains.*

11. Write the seven-letter word *element.*

12. Fe_2O_3 is ferric oxide, or "rust." Therefore, C + Fe_2O_3 = C + rust, or "crust."

13. Change the letter *m* to the letter *b* to spell *batter*—a flour mixture for cooking.

14. Mix up the letters of the words *iron* and *sand*: i + s + o + a + r + n + d + n.

15. Rock.

16. CO_2, which is carbon dioxide—a substance used in fire extinguishers.

17. Rearrange the letters in the word *phase* to spell *shape.*

18. Change the order of the letters in the word *physical*; for example, *pyshilac.*

19. Underline the letters *a, t, o, m,* and *s*: c<u>o</u>ns<u>ta</u>nt <u>m</u>o<u>t</u>i<u>o</u>n.

20.
```
        y        o
   h       r        e
        d       g        n
```

21. i<u>N</u>ert.

22. ELECTRONS

23. Remove *non* from the word *nonmetal.*

24. Use letters in the word *motion* to spell *tin*: mo<u>ti</u>o<u>n</u>.

25. The metallic elements Mo (molybdenum) and Mn (manganese).

26. Two or more are called *nuclei,* which fits into the spaces.

27. Levels *1, 2,* and *3* are all the same: the word *energy.*

28. Fish: F (fluorine), I (iodine), S (sulfur), H (hydrogen).

29. The word *he* makes up 20 percent of the word *alchemists.*

30. Solids: AL (aluminum), C (carbon). Gas: He (helium). Liquid: mist (water droplets).

31. Write the word *alchemist* twice, in the form of a cross:

```
            a
            l
            c
            h
    alchemist
            m
            i
            s
            t
```

32. Coal.

33. (a) rat, (b) ram, (c) mare, (d) tram, (e) meter, (f) mat, (g) arm, (h) tee, (i) tea, (j) tear.

34. Part A: gas. Part B: steam.

35. (a) boa, boar; (b) ram; (c) eel; (d) ant; (e) lion; (f) cat; (g) asp, rat; (h) hen; (i) emu; (j) hare.

36. (1) lithium, (2) neon, (3) sulfur, (4) hydrogen, (5) silicon, (6) helium.

37. Part A: (1) S, sulfur; (2) Mn, manganese; (3) Te, tellurium. Part B: chlorine.

38. (a) TeRn, (b) RaW, (c) WAtEr, (d) ICe, (e) BONeS, (f) FeRn, (g) NoBODy, (h) BrISK, (i) HeAr, (j) CoAt.

39. (a) log, solid; (b) neon, gas; (c) grain, solid; or rain, liquid; (d) air, gas; (e) arm, solid; (f) ice, solid; (g) cent, solid; (h) argon, gas; (i) mist, liquid; (j) hare, solid.

40. Iron.

41. Carbon.

42. Ice.

43. Seed.

44. Prism.

45. Coins or money.

46. Lemon.

47. Precipitation.

48. Giant panda.

49. Electron.

50. Chemical properties.

51. Sol: a name that refers to the sun, which produces gas.

52. Rope: pr<u>ope</u>rty.

53. A partial vacuum.

54. Article: <u>p</u>a<u>rticle</u>.

55. Plenty of lattice.

56. Fe + Y = iron + Y, or "irony."

57. Lid: so<i>lid</i>.

58. Remove the *i* in so*lid* and mark it so*ld!*

59. The word *oil* is 75 percent of the word *soil*. Gas comes from oil.

60. Mica: che<u>mica</u>l.

61. The "ez" part (of the word *freeze*.)

CHAPTER 2: MINERALS

62. Iron.

63. In a ring: occu<u>rring</u> naturally.

64. All three minerals fit the description of a solid: matter that has a definite shape and takes up a definite amount of space.

65. BeAr spells *bear*—a living creature.

66. A lite soda.

67. A *tourma*.

68. Four: three minerals and one "mineral": <u>Min</u> <u>E</u>. <u>Ralston</u>.

69. dia↓mond or ←diamond

70.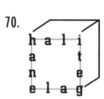

71. He placed both specimens on his desk. Then he shined a flashlight on the hematite. Thus, it appeared lighter.

72. Switch the *i* with the *l* to form *sliver*.

73. Gem.

74. Possible answer: cleavage → east.

75. Draw a crystal ball.

76.
```
 l            l
 e    l       e
galena      lead
 d    e       d
 a    a
 a    d
 d
```

77. Possible answer: A quantity of quality quarry quartz.

78. Possible answer: A mallet can mash a malleable mineral.

79. Possible answer: A cluster of colorless cubical crystals.

80. Use letters in the phrase *a cocoon* to write the chemical formula for calcite, $CaCO_3$.

81. Write the word *crystals* four times, from top to bottom, in the grid. Letter sizes will vary from large to small.

82. Rearrange the letters in *tin* to spell *nit*—a louse egg.

83. Rearrange the letters in *ore* to spell *roe*—which are fish eggs.

84. Use the large *A* and small *u* for the symbol of gold, Au.

85. Use the letters *si* in *silver* to spell the chemical symbol for silicon, Si.

86. Write the word *crystal*.

ANSWERS—Chapters 2 & 3

87.

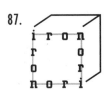

88.

99
S

89. Work the "mine" in mineral. Break up enough rock and you're bound to find several different minerals.

90. Lay the minerals on a soft bed in a wagon. Slowly pull the wagon and the minerals shouldn't break.

91. Pyrite: the pi sign for "py," + r + "i" + t + e.

92. (a) halite, (b) calcite, (c) talc, (d) limonite, (e) flint, (f) biotite, (g) sulfur, (h) water, (i) cuprite, (j) magnetite.

93. (a) corundum, (b) galena (galeNa), (c) H_2O (water).

94. A. graphite: (1) "hi," (2) ap, (3) gr, (4) te.
B. cinnabar: (1) ar, (2) nn, (3) ci, (4) ab.
C. pyrite: (1) py, (2) te, (3) ri.
D. galena: (1) le, (2) na, (3) ga.

95. Corundum; 53 percent.

96. Feldspar; orthoclase (the numbers represent the order of letters in the alphabet).

97. Copper.

98. Iceland spar.

99. Corundum.

100. Magnetite.

101. Barite.

102. Iceland spar.

103. There is no letter d in galena; there is only lea.

104. It's too old: gold.

105. In the form of a dime: sediment.

106. Fe in feldspar is the chemical name for iron.

107. He in Hematite is the chemical name for helium.

108. Two: minerals and minerals.

109. A mineral found in limestone and marble.

110. The letter O.

111. The center: asbestos.

112. A "Cu-Cu" clock.

113. Meg Rotcelloc ("gem collector" spelled backwards).

CHAPTER 3: ROCKS

114. Mixture of minerals.

115. Basalt: Ba + Na + Cl = (Ba + NaCl) = (Ba + salt) = Basalt.

116. Plutonic.

117. Each cavity has it in the middle: cavity.

118. Granite. Some granite is comprised of quartz, feldspar, mica, and hornblende.

119. Replace the letter t with an s in clastic and you'll have a "classic."

120. Change the l to a t and the h to an l to form slate.

121. He met-a-morphic.

122. A mess!

123. Coal—a sedimentary rock formed from plants, which are organic.

124. A rock cycle.

125. Lava flows in all directions. Therefore, write the word lava in the center spaces of the grid, and use arrows to show how lava spreads in all directions.

126.
```
                        i    a
                          r
heat ————————→    s  e  n
pressure ————→              l
                        m
```

127. <u>certain conditions</u>
     limestone

128. <u>pressure</u>
     marble

129. Replace tin with the chemical symbol for tin, Sb, to form serpenSbe.

130. Remove letters i, g, u and s. The letters neo remain, which means "new."

131. Heat transforms limestone into *marble*, which is a word with three less letters.

132. Use a rock saw to cut it open.

133. Use the middle letters from the word *concretion* as the nucleus:

134. Fine texture **COARSE TEXTURE**

135. Part A: gneiss, slate. Part B: (1) gnat, (2) lint, (3) gist.

136. Part A: (1) rain, (2) art, (3) tone, (4) con, (5) dime. Part B: diorite.

137. Pumice; half of the rock is ice (pum*ice*).

138. Shale.

139. Lava.

140. Obsidian.

141. Igneous.

142. Rock cycle.

143. Pumice.

144. Limestone.

145. Gneiss.

146. Tillite.

147. Slate.

148. It was under a lot of pressure at the time.

149. Two.

150. Mice: pu*mice*.

151. "You've got rocks in your head."

152. In a *bat*holith.

153. In shale beds.

154. Uplifting.

155. Mud pies.

156. Rock 'n Roll.

157. Organic donors.

158. Fossi*lifer*ous limestone.

159. When it becomes quartzite.

160. People found him too abrasive.

161. Liquid state.

162. An avenue: c*ave*.

CHAPTER 4: VOLCANOES

163. The letters *c* and *s* in the word *cones*.

164. Press the letters on a typewriter or keyboard that spell the phrase *a volcano*.

165. A big "gash."

166. The plural form *magmas* contains the letters needed to spell *gas*, whereas the singular form *magma* does not.

167. Possible answer:

```
        e
        r
     crust
        p
        t
        i
        o
        n
```

168. Spell the word backwards to form a "pure" eruption: noit*pure*.

169. Possible answer:

170. Invert the *V* from the second *volcano*.

171. Invert the *W* in the word *Hawaii*, and then slant the legs of the *M* to resemble a volcanic cone.

172. The word *backlash* means "a sudden or violent reaction." To produce *backlash* from *black ash*, rearrange the letters so that *l* follows *k*.

173. <u>Magma</u>.

174. Write *oNaClov*: NaCL is the chemical symbol for salt.

175. The word *sure* is found in *fissure*—an open fracture in the ground.

176. *Com* is part of the word *composite* as well as one-half of the word *common*.

177. Write the word *volcano*. Then erase every letter except the *l*: the erasing of the word represents erosion; the letter *l* represents the remaining volcanic neck.

178. Shape the two *m*'s in the word *magma* to look like volcanic cones: ⌒a g⌒a

179. Write the term twice: *volcanic eruption, volcanic eruption.*

180.
v o l c(a)n o
v o l c(a)n o

181. Cools, hardens, igneous rocks.

182. (a) alive, (b) sleeping, (c) dead.

183. Mount St. Helen's; earthquake (th + ke + qua + ear).

184. Haleakala; "House of the Sun."

185. A crater skater.

186. A spent vent.

187. Scaldera.

188. Two: v<u>o</u>lcan<u>o</u>.

189. The *no* part of the word *volcano*.

190. It lacks a *bath* in front of the *olith*.

191. An event: e-vent.

192. Either *vlao* or *vola* remains:

V Ø L Ȼ A N O Ȩ Ş

193. "Look at me. See what happens when you stick your neck out?"

194. Tsunami ("salami") on Rye.

CHAPTER 5: EARTHQUAKES

195. The "upper crust."

196. Smog: sei<u>smog</u>raph.

197. **epicenter**
a
r
t
h
q
u
a
k
e

198. It was nobody's "fault"; it just happened.

199. Use 80 percent of the letters in the word *earth* to spell *heat*.

200. Four possible answers: *epicen, picent, icente,* and *center.* (Two out of three stations means that 33 percent of the *epicenter* would be missing).

201. (a) rat, (b) ant, (c) roach, (d) bear, (e) sheep, (f) bat, (g) lamb, ant (h) ram.

202. Earthquakes can cause damage "to a lot."

203. Change the order of the letters in the word: *ctusr, cstru, rtcus,* and so on.

204. The symbols can be used to show volcanic activity:

205.

206. Write the word *earthquakes* six times.

207. Highlight the *vast* part of *devastation*: de<u>vast</u>ation.

208. d⌒i s⌒e a s⌒e s →

209. The letters in *scar*, *car*, and *carp* help form the word *scarp*.

210. About 50 percent of a *tsunami* is *sun*: ts<u>un</u>ami.

211. Disastrous; heat.

212. Part A: (1) fat, (2) ave., (3) epic, (4) nit, (5) ten.
     Part B: (1) tremor, (2) energy.

213. (1) focus, (2) ear, (3) rip, (4) eater, (5) pet, (6) sun, (7) ant, (8) tag, (9) eeg, (10) S-wave, (11) magma, (12) era, (13) pride, (14) map.
     Puzzle answer: 1906.

214. Seismograph. The race results: Parsley, 1st; Sagebrush, 2nd; and Lilac, 3rd. (Parsley begins with "P" for P-wave; Sagebrush, "S" for S-wave, and Lilac, "L" for L-wave. (P-waves travel faster than S-waves, which travel faster than L-waves.)

215. San Andreas; the two plates are called the Pacific and the North American.

216. Tsunami.

217. Fault.

218. Aftershock.

219. Earthquake.

220. Energy.

221. Both create destruction at the edge of a plate.

222. An epicenter.

223. *Q* and *R*—the two letters between *P* and *S* in the alphabet.

224. In a private "fault."

225. Rock and roll.

226. The "earth" and a "quake."

227. A golfer's reaction to a botched shot.

228. Because she was about to snap under the pressure.

229. Because they won't stay in "focus."

230. Register on the Richter scale.

231. With a seismic wave.

232. The "quake" part.

233. <u>J</u>ump<u>y</u>.

234. <u>Moo</u>dy.

235. <u>Fl</u>ighty.

236. <u>Buzz</u>-arre.

237. <u>Upro</u>ariously.

CHAPTER 6: FORCES ACTING ON THE EARTH'S CRUST

238. The final condition would be dust: Another name for *muscovite* is *mica*; delete *mica* from *dumicast* and only "dust" remains.

239. Mechanical weathering: no mineral change (xy).
     Chemical weathering: a new substance forms (wz).

240. Water, wind, and waves.

241.

242. Motion.

243. Heed.

244. There would be nothing "left," but there would be something "right": the letter *i* to the right of the vertical line.

245. Gully: (a) gull + (b) Y.

246. *Plateauru* was left: remove *SOS* from the word *Plateosaurus*.

247. They planned to "runoff" together.

248.

249. Foliate: foliat + e.

250. Moving air.

251. Possible answers:

```
        i c e   c r y s t a l s
        c                     l
        e                     a
                              t
ice crystals                  s
ice crystals        c         y
ice crystals  or    r         r
ice crystals        y         c
ice crystals        s
                    t         e
                    a         c
                    l
          s l a t s y r c   e c i
```

252.

```
            a          a
            ↘          ↘
d i a c t r o p h i c m
      ↗          ↗
      s          s
```

253. Write the word *same* by rearranging the letters in *mesa*.

254.

```
c o
   ⌐n t i
        ↘
          n e n t
```

255. Lightly write the word *hill*, and then erase part of the word.

256. Part A: (1) astro, (2) tic, (3) eau, (4) via, (5) cur.
Part B: Gondwana.

257. In many caverns, underground lakes are formed as groundwater moves through limestone.

258. Gravity; replace the *g* with an *e*, and unscramble the letters to spell *variety*.

259. Gravity.

260. They fold under pressure.

261. The tonic part.

262. Alluvial fans.

263. Because it carries only the finest particles.

264. Water.

265. Because without a *part*, there would be no *particle*.

266. In a mo<u>rain</u>e.

267. Going down the landslide.

CHAPTER 7: PREHISTORIC LIFE ON EARTH

268. Fossil: soil.

269. In the title: pal<u>eon</u>tologist.

270. Write one-half of the word *animal* (i.e., *ani, nim, ima,* and *mal*).

271. Pronouncing its scientific name.

272. The animal had been dead for over 170 million years; fossils can't fly.

273. They discovered that animals did not eat after they died and became fossils.

274. "House of Preserved Game."

275. At least half: *fos . . .*

276. He grouped them as invertebrates: the word part *ebrates* was inverted.

277. Trilobite: three *lobite* or *tri-lobite*.

278.
```
f o s s i l
      f o s s i l
      f o s s i l
f o s s i l
```

279. FOSSIL

280. Add an *s* before the *n* to form a *snail*, an animal that can become a fossil.

281. Ache: Cockro<u>ache</u>s.

282.
```
                    l
                  a
                a
              m
            m
          a
        a
      m
    m
```

283. Write three end marks or periods (i.e., *. . .*).

284. "Terrible Reptile."

285.

```
        s
        t
        a
        r
        f
        i
        s
        h
Paleozoic era
        s
```

286. Oz: Mes<u>oz</u>oic and Cen<u>oz</u>oic.

287. Plant and sponge: p + l + "ant"; s + "pong" + e.

288. Part A: (1) zebra, (2) lion, (3) bear, (4) horse, (5) rhino, (6) camel, (7) bison, (8) man.
Part B: boar.

289. Woolly mammoth; there are enough letters for nine "ice," two "frozen water," and one "hail"; ma moth (<u>mam</u>moth); write the two names in the form of a cross.

```
          w
          o
          o
          l
          l
          y
          m
  elephant
          m
          m
          o
          t
          h
```

290. He always left the wrong impression.

291. "I've had my fill of you."

292. Trilobite.

293. Some are petrified.

294. "You're well preserved for your age."

295. Forever, Amber.

296. Because they only do fossil impressions.

297. An ant: sc<u>ant</u>y.

298. Because they have a "halo": cep<u>halo</u>pod.

299. Because some of them had a wingspan of 2 1/2 feet (.75 meter).

300. Large predators.

301. Fossils.

CHAPTER 8: WEATHER AND CLIMATE

302.

303. By writing the letters *cy* you'll produce a "cy-clone."

304. Use the X's to cross out "air" in *prairie*. A tornado forms when layers of cold air trap warm air. Without "air," a tornado cannot form.

305. By writing the word *rain* several times. Thus, a true <u>rain</u>maker.

306. <u>Wh</u>eater.

307. <u>Legum</u>inous refers to bearing a legume. One half of the word *climates* is *lima*, which is a legume.

308. Many listeners thought Nancy predicted inclement weather on a daily basis. After several days of fair weather, they lost confidence in her.

309. Weather- and climate-related words include *troposphere, hot, poor,* and *tops*.

310. The hidden sentence: *Cumulus clouds are puffy.*

311. Hurricane: It sets off a major alarm, and offers a "big eye" of charm.
Tornado: It can pick up a tree and a farm, and will do it with no hand or arm.

312. cli + cli, which equals cli-"mates."

313. w + i + n + d + s.

314. The "ado" in *tornado* would be crunched. The "torn" part would damage the "ado."

315.

$$\frac{\text{fair} \longrightarrow \text{stormy}}{\text{daily basis}}$$

316. Possible answer:

```
  at
 mos
 phe
  re
```

317. Possible answer:

water ⟶ water
Texas ⟶ Missouri

ANSWERS—Chapters 8 & 9

318. Remove any two letters in *humidity*.

319. P<u>recipita</u>tion (rain).

320. (a) high, (b) blunder, (c) condensation, (d) smog, (e) gust, (f) gale.

321. Part A: two.
Part B: much rain.

322. (a) thermometer, (b) barometer, (c) anemometer.

323. Doldrums; "the great circle of the earth"; circle the letters e, q, u, a, t, o, and r to form *equator*.

324. Tundra; tin or "ten."

325. As a "wind, wind" situation.

326. Because they all end up . . . "here." (troposphere, stratosphere, and so on.)

327. Because it is given in knots.

328. Cirrus clouds.

329. They all end in "ation."

330. Because he put up too much of a front.

331. Because the fact that they forecast weather is always true (it's their job).

332. Because the forecaster chose to predict a shower rather than take one.

333. Mercury barometer.

334. They both do their share of "squalling."

CHAPTER 9: ASTRONOMY

335. MOON.

336. Turn the *V* in *Venus* upside down.

337. Underline or circle the word *pit* in *Jupiter*.

338. Six: Venus, Saturn, Mars, Uranus, Neptune, and planet.

339.
```
  m
  e       m
  t       em          m
  e       te          e
comet's  path
  r       oe          e
          ro          o
          r           r
```

340. Write *comet* backwards: t e m o c.

341. The name of the planet is *target*:

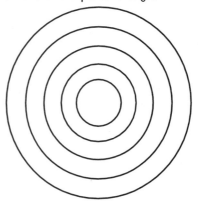

342. "Dirty" tub. The discovery: It had a ring around it.

343. A telescope.

344. Possible answers: head, eye, hand, ear, heart, teeth, neurons, throat, spleen, and ulna.

345. Five: Divide the figure in half and you'll have two extra *t*'s. Now there are enough letters for five complete spellings of *meteor*.

346. Place the words *night sky* in the middle of the phrase: the—night sky—day.

347. Clips: e<u>clips</u>e.

348. Underline and unscramble UFO: <u>FULL M</u>OON.

349. The earth's closest star is the sun. The word *sun* appears backwards in *Venus* and *Uranus*.

350. The word *rest* means "a lack of movement." Without the letters of the word *rest*, there would be no *terrestrial*.

351. Write three-fourths of the word *year*; for example, *yea* or *ear*.

352. Rearrange the letters of the word *star*; for example, *srat, trsa, rsat,* and *tsra*.

353. The letter *t*: cra<u>t</u>ers.

354. Stella: co<u>nstell</u>ation.

355. (1) Camelopardalis, (2) Gemini, (3) Auriga.

356. (1) jovian, (2) aphelion, (3) zodiac, (4) nebula, (5) planetoid.

357. Uranus; Herschel's first name is William; 84 Earth years equal 1 Uranus year.

358. Milky Way; the shape of the structure is a spiral.

359. Asteroids; there are about 1,000 in our solar system.

360. An "extra terrestrial."

361. There is a *ton* in *stony*; a ton is equal to 2,000 pounds.

362. They use the asteroid belt.

363. About 12 hours.

364. Because it goes through phases.

365. By striking the earth's surface, thus becoming a meteorite (and adding letters *ite*).

366. Two: M<u>OO</u>NS.

367. The crescent phase.

368. Between the *r* and the *y*: a *cur* is "a mongrel or inferior dog."

369. A partial eclipse: When you remove *lip* from the word *eclipse*, only part of the word remains.

CHAPTER 10: OCEANOGRAPHY

370. Take away NaCL (salt) from the word and *iniy* remains.

371. Layer.

372. The ocean current was responsible for pushing the bottle around for 20 years; thus, a "current event."

373. Chemically speaking, a diatom means "having two atoms."

374. The name *cormorant* is hidden in Carl's order: <u>c</u>old, <u>o</u>ily, <u>r</u>ed <u>m</u>eat <u>o</u>n <u>r</u>ye, <u>an</u>d <u>no</u> tomato.

375. A *scallop* has no *i*'s, but the animal can still see.

376. Abyss.

377. People would have to pay "a toll" to enter "atoll."

378.

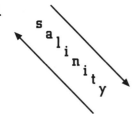

379.

life
ocean floor

380. The "plan" part: <u>plan</u>kton.

381. By pointing out how they become attached to rocks.

382. Change *algae* (plural) to *alga* (singular).

383. Fish, eel, and whale.

384. There are ten "tacle": Ten-tacle, or tentacle.

385. There are six spellings of *orca*: You need to use the circle enclosing the letters as a letter *o* to complete the spelling of the sixth *orca*.

386. (1) eel, (2) whale, (3) snail, (4) algae, (4) herring, (5) cod.

387. (a) wing, (b) yes, (c) no, because the four unused letters spell *wing*; it would be difficult for a bird to fly missing a wing.

388. (1) c, (2) d, (3) f, (4) a, (5) e, (6) b. (Sketches will vary.)

389. Jellyfish; USA (me<u>usa</u>).

390. Octopus: oc<u>top</u>us.

391. Abalone: ab<u>alone</u>.

392. Coral.

393. Tide.

394. "Swell" condition.

395. A sole survivor.

396. Seamount.

397. "Face it: You'll always be spineless."

398. A skydiver.

399. When they go from crest to trough.

400. Tide pool

401. Sea urchin: <u>urchin</u>.

alchemist—medieval chemical scientist

algae—aquatic plant-like organism

ammonia (NH_3)—colorless, gaseous compound of nitrogen and hydrogen

andesite—dark grayish rock consisting of oligoclase or feldspar

apatite—calcium phosphate mineral found in bones and teeth

Archeopteryx—prehistoric bird from the Jurassic period

asbestos—mineral that is inextinguishable when set on fire

atoll—circular coral reef surrounding a lagoon

barium (Ba)—silver-white, metallic, toxic, malleable element that occurs only in combination

barometer—instrument for determining atmospheric pressure; used for weather forecasts

basalt—dark gray or black dense igneous rock

batholith—great mass of igneous rock that stopped its rise below the earth's surface

bauxite—mineral that is the principal source of aluminum

benthos—plants and animals living on the ocean floor

biotite—black or dark green form of mica

Cactocrinus nodo brachiatus—sea lily

calcite ($CaCO_3$)—mineral consisting of crystallized calcium carbonate, limestone, chalk, and marble

caldera—volcanic crater formed by an explosion that creates a basin-like depression

carbon dioxide (CO_2)—heavy colorless gas formed in the respiration of animals

Cenozoic era—era of geological history marked by mammals rising from obscurity to the dominant position in the animal kingdom

cephalopods—class of marine mollusks (squid, octopus) with sucker-bearing arms; move by expelling water from a siphon under the head

Ceres—Roman goddess of agriculture

chlorine—element isolated as a greenish-yellow gas; often used as a bleach

cinder—hot coal without a flame

cinnabar—artificial red mercuric sulfide used as a pigment

clastics—group of sedimentary rocks composed of different-sized fragments

cleavage—the way minerals split apart

composite—made up of distinct parts

concretion—round objects or odd-shaped masses of cemented material that collect around a nucleus; feature of some sedimentary rocks

condensation—change of a substance (water) from vapor to liquid or solid

conglomerate—sedimentary rock composed of round pebbles cemented into a solid mass

constellation—configuration or arrangement of a group of stars

cormorant—large diving water bird with webbed toes and a hooked beak

corundum—hard mineral that occurs as colored crystals (ruby, sapphire)

crustacean—aquatic animal whose skeleton is on the outside (exoskeleton), as in lobsters, shrimp, and crabs

cuprite—copper ore

cur—mongrel or inferior dog

diastrophism—movement of the earth's crust that causes rock layers to fold

diatoms—one-celled water plants with walls of silica

diorite—granular crystalline igneous rock

downdrafts—strong vertical winds occurring during thunderstorms

ductile—capable of being pulled into thin strands without breaking

electroencephalogram—instrument used for detecting and recording brain waves

emu (electromagnetic unit)—system of electrical units based on the magnetic properties of electrical currents

epicenter—point on the earth's surface where an earthquake begins

epoch—memorable event or date

erosion—the breaking up and carrying away of material from the earth's crust

VOCABULARY LIST

evaporation—to dissipate or draw off in vapor or fumes

exfoliation—the peeling away of surface rock due to weathering

feldspar—any of a group of crystalline minerals; an essential part of nearly all crystalline rocks

fissure—open fracture around the base of a volcano

galena—lead ore

geodes—hollow rocks lined with crystals

glacier—body of ice that moves slowly down a valley or depression

gneiss—foliated metamorphic rock (like granite)

granite—hard, natural, igneous rock of visibly crystalline texture

guyots—submerged ocean peaks (seamounts) with flat tops

gypsum—common mineral used in soil and plaster of paris

halite—rock salt

hematite—iron ore

Homo sapiens—humankind

hornblende—common dark variety of minerals with like crystal structures

hypersthene—grayish, greenish-black, or dark brown igneous-rock-forming minerals

igneous rock—rock made from hot liquid matter that has cooled and hardened

Jurassic period—period marked by the presence of dinosaurs and the first appearance of birds

L-wave—"surface" or "longest" seismic wave; travels only on the surface of the earth and stretches out for miles

laccolith—mass of igneous rock intruded between sedimentary beds; produces dome-like bulging of overlying strata

landmass—large area of land

leguminous—of or bearing legumes (vegetables)

limestone—rock formed by the accumulation of organic remains; used extensively in building

limonite—major ore of iron

magma—molten (melted) rock within the earth

magnetite—black mineral; major ore of iron

Mesozoic era—era marked by the presence of dinosaurs, marine and flying reptiles, ferns, and the appearance of mammals and birds

metamorphic rock—formed when heat and pressure change one kind of rock into another

meteorite—meteor that reaches the earth's surface without being completely vaporized

meteorologist—scientist who studies the atmosphere and its phenomena; especially weather and weather forecasting

methane (CH_4)—colorless, odorless, flammable gas; product of decomposing organic matter

mica—any colored or transparent mineral crystallizing into forms easily separated into thin "leaves"

microcline—mineral of the feldspar group

-morphic—having a form (meta*morphic*)

mudflow—moving mass of mud created by rain or melting snow

muscovite—another name for *mica*

nektons—free-swimming aquatic animals

nimbus—storm clouds; thunderhead

nucleus—central core of an atom

obsidian—dark natural glass formed by the cooling of molten lava

P-wave (primary wave)—"compressional" or "push-pull" seismic wave; travels fastest and arrives first

paleontologist—scientist who studies prehistoric life and fossil remains

Paleozoic era—era of geological history marked by the appearance of nearly all invertebrates except insects, and later, by the appearance of terrestrial plants, amphibians, and reptiles

Pangaea—name given by Alfred Wegener to the landmass he theorized existed 200 million years ago, which consisted of all the continents

pegmatite—coarse variety of granite

pelagic—describes anything related to or living just above the ocean floor

Pennsylvanian period—period in the Paleozoic era marked by the earliest appearance of reptiles

petrologist—scientist who studies the origin, history, occurrence, structure, chemical composition, and classification of rocks

plankton—tiny floating organisms

Plate Tectonics theory—geological theory in which the earth's crust is divided into several "plates" that float on and travel independently over the mantle

Plateosaurus—dinosaur that lived during the late Triassic period (about 220 million years ago); grew to about 20 feet (6 meters) long, and had a long neck and small head; first common plant-eating dinosaur

Pleiades—cluster of stars in the constellation *Taurus* that includes six stars formed as a "small dipper"

Pleistocene epoch—part of the Cenozoic era marked by the appearance of the earliest humans

Portuguese man-of-war—large, tropical, free-swimming/floating animal of delicate and transparent nature; uses bladder-like float and possesses tentacles

precipitation—deposit of water on the earth in the form of hail, mist, rain, sleet, or snow

pumice—lightweight volcanic glass with cavities; used in powder form for polishing

quartz—mineral appearing as colorless and transparent or as colored hexagonal crystals

rainmaker—person who attempts to produce rain through artificial means

rhyolite—acid volcanic rock that is the lava form of granite

Richter scale—instrument used to measure the magnitude (amount of energy released) of an earthquake

S-wave (secondary wave)—"shear" or "shake" seismic wave; travels slower and arrives later than the P-wave

salinity—amount of salt and dissolved minerals in water

sandstone—sedimentary rock consisting of cemented grains of quartz sand

scarp—small cliff formed by erosion

schist—metamorphic crystalline rock with closely foliated structure; easily split along parallel planes

seamount—submerged ocean peak

sediment—matter that settles at the bottom of a liquid

sedimentary rock—rock formed from accumulated layers of sedimentation in rivers, lakes, and oceans

seismic—of or caused by an earthquake

seismogram—record of an earth tremor by a seismograph

seismograph—instrument used to measure and record vibrations within the earth and on the ground

seismologist—scientist who studies earthquakes

serpentine—non-banded metamorphic rock

shale—rock formed by the consolidation of clay, mud, or silt

silicon (Si)—non-metallic element; most abundant element next to oxygen in the earth's crust

slate—dense, fine-grained, metamorphic rock made by the compression of sediments (clay, shale)

sodalite—transparent mineral found in igneous rocks

sodium (Na)—silver-white, waxy, ductile element; very chemically active

stratosphere—part of the earth's atmosphere in which temperature increases and clouds rarely form

subduction—process of the edge of one crustal plate descending below the edge of another

sulfur (S)—non-metallic element occurring in sulfides and sulfates; used in chemical and paper industries

syenite—igneous rock composed mainly of feldspar

talc—soft mineral with a soapy feel; used in making talcum powder

talus—slope formed by an accumulation of rock debris

till—glacial drift consisting of clay, sand, gravel, and boulders

tillite—mixture of boulders, sand, and clay "dropped" by a melting glacier

topaz—yellow quartz

tourmaline—mineral of variable color; makes a beautiful gem when transparent and cut

troposphere—lowest layer of the earth's atmosphere

Vesta—Roman goddess of the hearth

woolly mammoth—heavy-coated animal from the colder parts of the northern hemisphere

zeolite—any of various "watery" silicates with composition similar to feldspar; found in lava cavities